D0498045

The Peaceable Kingdom

BY JOHN SEDGWICK

Night Vision: Confessions of Gil Lewis, Private Eye (1982)

Rich Kids: America's Young Heirs and Heiresses,
How They Love and Hate Their Money (1985)

The Peaceable Kingdom

A YEAR IN THE LIFE
OF AMERICA'S OLDEST ZOO

John Sedgwick

WILLIAM MORROW AND COMPANY, INC.,
New York

The excerpt from *The Outermost House* by Henry Beston
used by permission of Henry Holt and Company.
Excerpts from *Wolf!* and photographs of the wolves at
the Philadelphia Zoo used by the kind permission of
Janet Lidle.
Photographs by Anne Bettendorf, Lauren J. Lewis,
Terry McBride, and Daniel Troy used by permission of
Asterisk, Inc.
Photographs from the archives of the Philadelphia Zoo
used by the kind permission of the Zoological Society of
Philadelphia.
Excerpts from *Animals Are My Hobby* by Gertrude Davies
Lintz, copyright 1942 by Robert M. McBride & Company,
pages 25–34.

Copyright © 1988 by John Sedgwick

All rights reserved. No part of this book may be
reproduced or utilized in any form or by any means,
electronic or mechanical, including photocopying,
recording or by any information storage and retrieval
system, without permission in writing from the Publisher.
Inquiries should be addressed to Permissions Department,
William Morrow and Company, Inc., 105 Madison Ave.,
New York, N.Y. 10016.

Library of Congress Cataloging-in-Publication Data

Sedgwick, John, 1954–
The peaceable kingdom : a year in the life of America's
oldest zoo
John Sedgwick.
p. cm.
ISBN 0-688-06367-5
1. Zoo animals. 2. Philadelphia Zoological Garden. I. Title.
QL77.5.S38   1988
590′.74′774811—dc19   87-24491
CIP
Printed in the United States of America
First Edition
1 2 3 4 5 6 7 8 9 10
BOOK DESIGN BY ELLEN FOOS

For Sara, who liked the elephants best

— PREFACE —

When you pass through the wrought-iron gates of the Garden of the Zoological Society of Philadelphia, you enter another world. The street noise fades, the city skyline drops away, and an extraordinary landscape stretches out before you. Tall trees sway, flowers blossom, and bright high-Victorian buildings loom. And everywhere there are animals—wild, uproarious, colorful animals. Llamas, camels, gibbons, polar bears, eagles, zebras, aardvarks, giraffes, rhinos, hippos, gorillas, peacocks . . . It is a kind of paradise, this place.

Like many people, I grew up with pets. Looking over a photograph album of my childhood recently, I was struck by how many dogs had passed through my life. First a pair of golden retrievers, then a line of Irish terriers, finally a series of mongrels my father preferred to describe as dogs "of uncertain ancestry" that we salvaged from the pound. I loved them all. As the youngest of three children, I paired off with the dogs, and I was happiest, most carefree, racing about the lawn with an animal barking at my heels. In my teens my family discovered the many charms of cats when a stray calico I named Esther wandered onto our porch. My father took to her especially.

But when I left home, I left the family pets behind. I might have acquired a dog or cat of my own, I suppose, but it didn't seem fair to keep an animal in a small apartment in the city. Without my realizing it, though, a vacancy had opened up in my life where these animals had been, and it was one that wasn't filled until I arrived at the zoo.

I came to the zoo to get away from the real world, as Walt Whitman expressed it, to "turn and live with animals" for a while in hopes that I might, like Whitman, be renewed by their calmness and simplicity. I had finished writing a book on the lives of young heirs and heiresses, and I was fatigued and a little depressed by the money culture. I was eager to get in touch with something more fundamental than money, to gain a fresh perspective on the world by seeing it, insofar as I could, from the animals' more basic point of view.

I was also keen to find people who were committed to their jobs not for the paycheck or the status (although working with animals is a mark of some celebrity), but for the sheer pleasure of doing something that is valuable in itself. And finally, I hoped to rediscover the rapture of being with animals that I had missed since my childhood. I thought, in short, that the Philadelphia Zoo would show me *life*.

Why the Philadelphia Zoo? Because it is America's oldest zoo, and because, to me, it is America's most intimate and charming. Other zoos—the San Diego, the National, and the Bronx—may be more famous, but I early sensed that their spreading fame had begun to constrain them. They acted more like image-conscious corporations and less like fun-filled zoos. Despite its impressive history, the Philadelphia Zoo is much less pretentious and much more inviting. As soon as I visited it, I knew I wanted to come back and stay a while.

I spent a year there, off and on, commuting from my home in Boston, starting in July of 1985 and ending the following June. My visits encompassed the four seasons, and gave me a complete view of life in this unusual ecosystem. I was especially interested in the relationship between man and animal, and I got to see it in all of its forms as it involved the chairman of the board all the way down to the lowliest keeper, and touched all the animals from the camels by the main gate to the wolves at the farthest corner. Except in one small case, I have changed no names, and I have reported everything precisely as I witnessed it. This book is the story of my year with the animals.

— ACKNOWLEDGMENTS —

First of all, I want to express my appreciation to the Philadelphia Zoo's president, Bill Donaldson, who so graciously invited me to come to the zoo and then let me loose to go wherever I wanted. I am also grateful to the zoo's public relations staff, Debbie Derrickson and Arlene Kut, who helped to facilitate my many visits.

Beyond that, I would like to thank all the members of the Philadelphia Zoo staff for their patience in answering my many questions about their lives and work. While I can't thank everyone, I would be remiss if I didn't single out veterinarians Keith Hinshaw and Mike Barrie; veterinary technicians Eileen Gallagher and Ann Hess; animal services personnel Dave Wood, Bill Maloney, Gene Pfeffer, Chuck Ripka, Dan Maloney, Al Porta, Bob Berghaier, and Roseann Giambro; Penrose Laboratory director Dr. Robert Snyder, curator Larry Shelton, and business manager Rick Biddle.

I am deeply indebted to my hosts in Philadelphia, Michael Kimmelman and Maria Simpson, and Nella and Bill Helm. Sue Quinn, Phil Zuckerman, and Corby Kummer offered much useful advice about the manuscript. Carolyn Lastick at the Philadelphia Zoo was a great help in assembling the photographs. Once again, my editor, Pat Golbitz, has served me admirably, as has my loyal agent, John Brockman. But finally and most important, I want to thank my family. My wife, the writer Megan Marshall, and our daughter, Sara, put up with my many absences with great understanding. And once a solid draft of this book was written, Megan went over it with the care and attention that only a fellow writer can provide. For this, and for so much else besides, thank you, my love.

— CONTENTS —

We need another and wiser and perhaps a more mystical concept of animals. Removed from universal nature, and living by complicated artifice, man in civilization surveys the creature through the glass of his knowledge and sees thereby a feather magnified and the whole image in distortion. We patronize them for their incompleteness, for their tragic fate of having taken form so far below ourselves. And therein we err, and greatly err. For the animal shall not be measured by man. In a world older and more complete than ours they move finished and complete, gifted with extensions of the senses we have lost or never attained, living by voices we shall never hear. They are not brethren, they are not underlings; they are other nations, caught with ourselves in the net of life and time, fellow prisoners of the splendor and travail of the earth.

—HENRY BESTON, *The Outermost House*

The Peaceable Kingdom

Prologue

SO. ST. PAUL PUBLIC LIBRARY
106 3RD AVE. N.
SO. ST. PAUL, MN 55075

Massa died the winter before I arrived at the Philadelphia Zoo. By far the most popular animal in the collection, he was at fifty-four the oldest gorilla in captivity, and quite likely the oldest gorilla that had ever lived. When he died, a darkness settled over the zoo that took some time to lift.

The night watchman found the 175-pound, gray-haired Massa slumped on his side, unmoving, in his center cage in the Rare Mammal House shortly before midnight on December 30, 1984. The zoo had just celebrated Massa's birthday that afternoon. Five hundred people had shown up to shower him with birthday cards and presents—bananas, mostly. The zoo commissary had concocted a huge cake consisting of a pint of vanilla ice cream, four bananas, four apples, four oranges, a bunch of grapes, and a quart of strawberries, all slathered with whipped cream and heaped on a chunk of "Zoo Cake," a heavy granolalike mixture of grains and vegetables the zoo serves its omnivores. Massa's longtime keeper, Ralph McCarthy, slid the cake under the bars while everyone sang "Happy Birthday."

Massa had been in decline in recent years, but he perked up for the party. He cavorted about his cage as he had of old, swinging from the steel bars and smacking the tire that dangled down into the middle of his cage with new enthusiasm. He enjoyed the grapes on the birthday cake particularly.

But now Massa was down in his cage and still. In a panic, the night watchman called head keeper Bill Maloney, who rushed to the zoo from his home in South Philadelphia. At Massa's cage, he reached in to prod the great beast with a stick. Maloney knew better than to go into the cage himself, or even to reach in with his hand. Gorillas can be dangerous. He well recalled the experience of an old keeper named Sammy Guanato who, despite Maloney's repeated warnings, used to hand peanuts through the bars to Massa's old gorilla colleague, Bamboo. One day Bamboo did not confine himself to Guanato's peanuts, but grabbed his whole arm and gave it such a yank that he plucked it right off at the shoulder. (Remarkably, Guanato was not fazed by the experience, but upon his return from the hospital, went right back to Bamboo's cage and waved his freshly stitched stump at the gorilla. "You didn't mean it!" he yelled. "I know you didn't mean it!")

Poking Massa with his stick, Maloney got no response. He called in the head veterinarian, Keith Hinshaw, who cautiously ventured into Massa's cage. He bent down beside the great ape to feel for a heartbeat, then ran an EKG to be sure. At about one o'clock on the morning of December 31, he pronounced Massa dead.

Massa's body was carried over to the nearby Penrose Laboratory on the zoo grounds, where pathologist Dr. Robert Snyder began his investigation into the cause of death. Despite the lateness of the hour, staffers started filing in to pay their respects. Debbie Derrickson, the young PR director, had sensed something terrible was happening, and she called in to the zoo from a restaurant. She was, consequently, one of the first to arrive. It was the oddest thing to see Massa stretched out, glassy-eyed and stiff, on the steel operating table. Death made him look more human than ever, she thought. Normally, the administrative staff is inured to news of a death in the animal ranks (to the animal services personnel, it can be another matter), but Massa was different. Like many people at the zoo, Derrickson had developed intense feelings for Massa. She used to go over to see him when job pressures were getting to her, the way one might visit a kindly uncle. But another thing was bothering her: the idea, soon put forth in the newspapers, that Massa's birthday party had killed him, for she herself had shopped for the whipped cream. As one headline put it: BEAST FEAST AND REST IN PEACE. "Oh great," Debbie thought, looking down at Massa, "the world's oldest gorilla, and I killed him."

As Snyder deftly sliced Massa open to look through his organs, he suffered no such qualms. What he discovered put Derrickson's mind at rest, for it turned out that Massa had been suffering from atherosclerosis, or narrowing of the arteries, for years. From a glance at the condition of the arteries leading from Massa's heart and kidneys, Snyder could see that the gorilla had suffered two heart attacks in the last month. A series of strokes that evening had finished him off. "This is a disease that has been going on for years," he told the press, "perhaps the greater part of a lifetime. Suddenly, one day, it's severe enough to cause a blockage of an artery. It doesn't really need a trigger."

In the days that followed, an unusual gloominess descended on the zoo. Some mourned openly. "But nobody should apologize for crying," said the gorilla keeper, Roseann Giambro. "It's okay to love. When people see their animals die, sometimes they say, 'I'm never going to get involved with an animal again.' But I don't see it that way. Love isn't *wasted*." Co-workers often called each other to make sure they were feeling okay. Even the zoo's portly director, Bill Donaldson, who rarely lets anything interfere with his good mood, found himself feeling sorry for the old ape. Born the same year as Massa, Donaldson thought of him as a Zen monk, and half expected that the gorilla would live forever. Like Debbie Derrickson, most of the workers at the zoo regarded Massa as a relative, and would take time out to commune with him at odd moments during the day. Dave Wood, a tough young elephant keeper who had recently taken over as the senior keeper's assistant, used to sit with Massa, holding hands through the bars. "Some people might think that was stupid, or risky, but to me it was worth it," he said later. "We had a really good relationship. I liked him and I think he liked me. He let me scratch his cheek. He'd make a low, rumbling sound, like he was purring."

Massa's keeper, Ralph McCarthy, never did get over the

gorilla's death. "I feel like I lost my best friend," he said. He used to call out, "Yo, Mat Mat," and Massa would come trundling over to him in what Ralph called his "Jackie Gleason shuffle" and then wheel around so that Ralph could scratch his back through the bars of his cage. Occasionally, Ralph would slip Massa a gumdrop or a pretzel. When Ralph's wife died a few months later, he felt that his whole world had crumbled. He took early retirement and left the zoo.

The Philadelphia Zoo never regarded Massa as just a gorilla but, then, Massa never saw himself as one either. He thought he was human. It had to do with the way he was raised.

Born in the Belgian Congo around 1930, Massa was bought as an infant from some tribesmen by a Captain Arthur Phillips who plied the African trade in his freighter, the *West Key Bar*. It's hard to know precisely what happened, but Massa's mother may have been killed while raiding plantation crops. If so, the Africans may well have spared Massa, thinking that gorilla babies fetched a good price from sea merchants. To keep him alive, an African woman may even have nursed him at her breast; it was a common practice. This might have been the start of Massa's confusion.

Captain Phillips had traded in such exotics before, and he brought the baby gorilla back to New York City, along with six chimpanzees, in the *West Key Bar*'s hold. He had previously sold some chimpanzees to a Brooklyn lady named Gertrude Lintz; when he reached New York, Captain Phillips called on her again.

A heavy-set Englishwoman, Mrs. Lintz was the wife of a wealthy and endlessly indulgent physician. The couple was childless, and perhaps as a substitute, Mrs. Lintz had developed an attachment to animals that carried beyond the bounds of a charming English habit to the point of definite peculiarity. As a young woman, she had started with a collection of prize St. Bernards that was rivaled only by that of Colonel Jacob Ruppert, Jr., the beer baron and onetime owner of the New York Yankees. But in 1914 her entire dog collection was mysteriously killed off by an unknown malefactor. Gertrude herself suffered a nervous breakdown, which put her in the care of an internist named Dr. William Lintz. He nursed her back to health; she married him. Recharged, she returned to raising St. Bernards and then branched out to less accommodating animals. First she trained some hummingbirds to eat out of her hand, then she made house pets of the tree squirrels in the backyard of their rambling estate on Shore Road, Brooklyn. Then, in quick succession, she raised tumbler pigeons, rex rabbits, a couple of ferocious giant horned owls named Jack and Jill, a leopard who was quickly remanded to a zoo when she jumped Mrs. Lintz and nearly cut her to pieces, and an Asiatic land lizard named That Devil that she never did identify properly.

But Mrs. Lintz discovered her life's work when a friend dropped by with a present of a tiny orphaned chimpanzee named Maggie Klein, who proved to be so clingy, adorable, and desperate for attention that Mrs. Lintz couldn't help but love her with her whole heart. So pleased was she with Maggie that she soon added another chimp, Joe Mendi. She taught the two of them to dress themselves. Maggie wore dresses, Joe Mendi a sailor suit. She took him shopping for shoes on Fifth Avenue and delighted in the looks of incomprehension from the salesclerks when she nonchalantly explained that young Joe was "a little ape, just over from Africa." Maggie also learned, after much practice, to thread a needle. Joe could hammer and saw. He could take nails out with the hammer, too, if he didn't cheat and use his fingers. Both chimps smoked.

Diverting as the chimps were, Mrs. Lintz had been longing for a gorilla ever since she had gone to the Ringling Brothers circus in 1915 and encountered a baby gorilla that had turned very sick. She imagined that in its homesickness it was "grieving itself to death," as she wrote in her memoirs, *Animals Are My Hobby*. "I had the feeling that if only I could take that lonely little gorilla I could love it back to health." She begged the authorities to let her try, but to no avail.

When Captain Phillips brought Massa to her in a cardboard box, she regarded it as an act of providence. Locked away for weeks in the damp hold of the *West Key Bar*, Massa had contracted double pneumonia and fallen unconscious. Mrs. Lintz leaped at the chance to nurse him back to health. For days on end she cradled him in her arms and fed him from a medicine dropper. Finally, the crisis passed, but the gorilla was still far from healthy. He was so feeble that Mrs. Lintz had to chew his food for him, softening it into a pap, and then place it in his mouth to swallow. This continued for months until one day she received some hothouse grapes in the mail. Seeing his mistress's excitement over the fruit, Massa reached out for the grapes and began feeding himself.

Mrs. Lintz named her baby Massa, the pidgin word for "Big Boss." Despite the name's masculine overtones, she thought her gorilla was female. It was an understandable mistake. Since gorilla penises are almost embarrassingly small, they are hard to discern from a distance, and gorillas don't generally welcome a close inspection. However, there does appear to be a certain willfulness on Mrs. Lintz's part to view her darling little Massa as female. In her memoirs she continues to refer to him as "she" long after the truth about the gorilla's gender had emerged. She dressed Massa in girl's clothes and claimed he was "the most feminine creature in the house, not excepting Missy." (Missy was the pet name she gave herself.) More likely, though, what Mrs. Lintz took to be Massa's femininity was merely a mimicking of her own behavior. One afternoon when the two were out driving, one of the gorilla's special pleasures, Massa took Mrs. Lintz's makeup kit from her handbag and powdered his nose so liberally it looked "as if she had been dipped in a barrel of flour." Mrs. Lintz also suspected that Massa girlishly envied her wardrobe. Several times she found him in her bedroom closet trying on some of her clothes. "I would come upon her with rage in my heart," Mrs. Lintz wrote, "and then collapse with laughter at the sight of Massa, lolling among the frivolous pillows of the chaise longue in borrowed finery, trying to look like Madame de Récamier." The gorilla would shuffle around the bedroom in Mrs. Lintz's shoes, drape her dresses about his body, and cram her hats on his head at unusual angles.

It is easy to find flaws in Mrs. Lintz's gorilla-raising techniques, but one has to remember that at the time, Massa was only the third gorilla ever kept in captivity in the United States. Given the prevailing attitudes of the day, it is remarkable that Mrs. Lintz even dared to try to keep one. Ever since Paul du Chaillu roamed about darkest Africa in the nineteenth century, sending back romantic and terrifying travelogues like *Sto-*

ries of the Gorilla Country filled with tales of killer apes, most people assumed that gorillas really were frightful murderers. The movie *King Kong,* which ruined gorillas' reputations for yet another generation, came out in 1933, two years after Mrs. Lintz acquired Massa. The image was not successfully countered until the 1950s and 1960s when the field studies of George Schaller and Dian Fossey showed that gorillas were in fact peaceable, rather fainthearted creatures that were much more likely to run from humans than to threaten them.

In any case, primitive as her methods were, Mrs. Lintz was undeniably successful, judging by her apes' longevity, one key measure that zoos use to ascertain their own effectiveness.

Soon Mrs. Lintz's friend Captain Phillips gave her another call, and she was put to the test with a second gorilla. This one she called Buddha, or Buddy for short. Under the name of Gargantua the Great, he grew to become the most famous circus animal of his day. At the time, though, he didn't look very impressive. He had been horribly disfigured by a sailor who, angry to have been sacked by Captain Phillips, took his revenge on the captain's most precious cargo. He sprayed Buddy with nitric acid from a fire extinguisher. His head and chest were horribly burned, his mouth misshapen, and he seemed to be blind. In his agony, Buddy refused to come out of the dark hold. But Mrs. Lintz went down with a flashlight and found a lonely figure cowering in the shadows. She lured him out with soft words and brought him home in a box.

An oculist restored Buddy's sight with eye drops, and a dermatologist performed crude plastic surgery that left Buddy's face contorted in the permanent scowl that would later create a sensation with circus audiences around the world. Buddy took to Massa immediately, although Mrs. Lintz gives little evidence of his sentiments being reciprocated, and the two of them went foraging together in the backyard, pouncing on robins and caterpillars. Buddy once picked Massa up in a gesture that Mrs. Lintz took to be romantic, and put him on the shelf in his bedroom. They went for rides together on the Lintzes' tea wagon. While Buddy did not hesitate to rebuke the smaller Massa with a cuff on the cheek, he could also be quite affectionate, and Mrs. Lintz says she sometimes found the two napping together, Massa's head pillowed on Buddy's belly.

With the addition of Buddy, the Lintzes' troop of great apes came to thirteen, the other eleven being chimpanzees. Mrs. Lintz ran her menagerie like a nursery school, and a strict English nursery school at that. To establish the basic standards of civilization, she dressed them all in human clothes, and she housed them in private rooms in a separate dormitory she called Simian House. While the gorillas, being a bit clumsier than the chimps, were exempted from some of the house rules, the apes were generally expected to make their beds, brush their teeth, and dress themselves. Maggie stood at the head of the class, since she could lace up her shoes and tie her sunbonnet in a bow. Eventually, Joe Mendi would learn to shave himself. Wearing bibs, they sat in high chairs at the dinner table for a breakfast of fruit, cereal, eggs, and milk. They ate with spoons and drank from cups. After breakfast they were taught how to walk erect. The older "children" had classes in typing, table setting, and operating the record player. Mrs. Lintz tried to teach her apes to talk, but didn't get very far. She could train them to follow some basic instructions like, "Give Missy the orange peelings." And Massa learned to warble along with her as she sang him lullabies in a rocking chair.

All the apes were toilet trained. Joe Mendi apparently was a fanatic about hygiene, but, as Mrs. Lintz writes, "in general gorillas and chimps [share] human disgusts." She explains, "If one of them happened to step in something unpleasant he would go lame in that foot, and limp with a horrified expression to find a bit of paper or rag to clean the offending member."

They napped at noon and played some hide-and-seek and tag afterward in the yard, but preferred the swings and jungle gym. Buddy boxed and wrestled with Joe Mendi. Massa shyly stayed out of the rougher games, but did consent to play a little tag with his Missy. Whenever fights broke out among the apes, Mrs. Lintz brought out the Bogeyman, a scary-looking Chinese head mounted on a stick that she taught them to fear by trembling before it herself. All the apes would immediately abandon their roughhousing and run for their lives when she produced it. At bedtime in the evening, Mrs. Lintz read them children's stories until they got sleepy. Then it was into their rooms for the night.

* * *

Of all the animals, Mrs. Lintz says she loved Massa best. "Massa is the most splendid being I ever knew," she writes. "For sheer sweep of temperament no ape, and no human being, has equaled Massa in my experience. So many pictures—Massa the delicate, helpless baby, needing protection even from the chimpanzees. Then the cuddly, charming little thing, begging for another ride on the tea wagon. Then the vain, feminine little girl dressing up in my clothes, preening herself . . ."

But her joy would turn to heartbreak and nearly to tragedy. Thinking that gorillas might have a future as domestics, Mrs. Lintz tried Massa out with the mopping. She immediately encountered problems. "Massa was so powerful that she could scrub the linoleum off the floor," she noted. But she persisted in the notion nonetheless—until one morning. Massa was down on all fours mopping the floor when Mrs. Lintz came into the kitchen unexpectedly. She slipped on a soapy patch and flew into the air, then landed with one foot on the pail of water. The pail overturned, and Massa was drenched. Suddenly, the domestic went wild. Mrs. Lintz writes:

[Massa] whirled about and rose to her full height, face contorted, eyes closed. There came from her a roar like nothing on this earth, a frightful, tremendous, howling screech as if the inferno had opened up to express in one sound all its dark fury. One cannot hear such a sound; it is not meant for us. It took from me all power to struggle or save myself, and yet a corner of my mind was telling me, this is the warning when the gorilla is ready for the kill. Now the beast will charge me—

In another instant she was rushing at me, jaw pushed forward to show her powerful fangs; and my own instincts came alive. I must protect my throat. I threw an arm across it as she sank her great teeth into my thigh, then again and again into my abdomen, ripping the flesh like paper. Then the beast made for my throat, catching my arm in her teeth. I tore it free, and the beast caught it again . . . over and over, how many times? I was conscious that an artery was torn, and bright jets of blood were spurting on the wall behind my threshing arm.

But this wasn't a beast, it was Massa. I must remind her, quickly, in Missy's voice—

"Massa, Massa! This is your Missy!"

The beast didn't hear me. But in the next room my brave little friend [a young houseguest] heard my voice and realized that I was in the room with the creature who sixty seconds before had uttered that dreadful roar. She rushed in, seized a heavy iron skillet from the stove, and with all her strength brought it down on Massa's skull.

Massa was stunned for a moment, so I could turn over and press my weight down on the torn artery. I began crawling away, and felt Massa's teeth in my leg again, but I kicked with the last of my strength. As the darkness came down, I knew I was safely away.

It took over seventy stitches to close all her wounds. As she lay in bed recovering, she worried about Massa. Had he meant it? Had he changed? One night she slipped out of her bed and went into his cage to see him. He greeted her warmly, with chuckles and a tender embrace. To determine her standing with the gorilla, Mrs. Lintz decided to test him with a simple command and asked him to sit in his chair. Always before, Massa would have meekly obeyed. Now he growled defiantly. Mrs. Lintz repeated her request in a firmer tone, but Massa continued to snarl. Mrs. Lintz backed off, and sadly returned to her bed. Having asserted himself, Massa was not about to return to docility.

Mindful of Massa's temper, Mrs. Lintz decided to confine him in a basement cage, and assigned her workhand, Dick Kroener, to watch over him. Massa never warmed to Dick, but he continued to act nicely toward his mistress, who always brought him sweets and amusements. She continued to enter his cage freely. One Sunday evening when Dick was out, Mrs. Lintz came into his cage bearing blankets for bedtime; this time, Massa was not so well behaved. He burst past her through the open cage door, dashed into Dick's bedroom, and trashed the place. He tossed everything in a heap in the center of the room, then tore Dick's bed to bits for good measure.

The Lintzes called the police, who quickly appeared with revolvers drawn. Mrs. Lintz pleaded with them not to shoot until she gave the word. Massa marched from Dick's room to the billiard room and was poised to demolish that as well when the Lintzes' African gray parrot, Charlie McCarthy, spoke up in a voice that sounded so exactly like Mrs. Lintz's that Massa stopped in his tracks and gazed up quizzically at the bird. As the police and the Lintzes looked on dumbfounded, the young gorilla retired to the rocking chair to think things over; then he went to the kitchen for a beer. After he had drained his third bottle, Mrs. Lintz felt that she'd better intercede, so she grabbed a chair and, brandishing it like a lion tamer, she forced the gorilla back to his cage. The police were impressed.

That did it, for Massa. Mrs. Lintz had no choice but to put him up for sale, and the Philadelphia Zoo won the bidding with an offer of $6,000. Like Mrs. Lintz, the zoo was under the impression that Massa was female, and it was eager to have a mate for its impressive male, Bamboo.

Mrs. Lintz drove Massa to his new home in the family station wagon, arriving just after Christmas, December 30, 1935, the day that the zoo would use afterward to celebrate Massa's birthday. When Massa was first displayed there to a team of researchers that had gathered to look over the new animal, he beat his chest ferociously. Mrs. Lintz was so proud of him she burst into tears. She stayed on in Philadelphia for three weeks to ensure Massa's smooth transition into his new home, then she drove alone back to Brooklyn.

At the zoo Massa was placed in the cage next to "her" gorilla fiancé in the massive red-brick Monkey House. Nine years old, Bamboo was then a hefty 340 pounds. Only five, and always on the skinny side, Massa weighed a mere 150. To the zoo staff, Massa's lean figure confirmed their belief that they had a comely female on their hands.

The newspapers gleefully played up this love match. The Philadelphia *Record* ran a formal announcement of Miss Massa's betrothal to Mr. Bamboo on the society page as if it were a Main Line event. And all the papers sent out feature writers to record the prenuptial arrangements. They wrote up charming accounts of Bamboo beating his chest and "roaring his jungle lovesong" to his blushing bride and of Massa responding coquettishly with mild sashays of his own. Massa would playfully toss some straw over at Bamboo to get his attention and then, as he used to in Mrs. Lintz's bedroom, cavort about his cage with a dress over his head.

But, for all the devoted wooing, when the two were finally introduced that August, they proved to have irreconcilable differences, or more accurately, irreconcilable similarities. Massa reacted to Bamboo's advances in a most unladylike fashion. He strode into their common cage and knocked Bamboo over with a roundhouse right. The two gorillas then "went into a clinch," as one paper put it. As they separated, Massa yanked Bamboo's chest hair, popped him one more in the face, and then scampered off through a passageway too small for Bamboo to follow. Safely back in his cage, he made faces at the bewildered Bamboo through the bars. Massa repeated these sudden gorilla raids seven times before the zoo's curator of mammals, Frederick Ulmer, halted the proceedings.

Zoo officials must have wondered what they had on their hands and formed a committee to investigate the delicate matter of Massa's gender. A few days later, the committeemen delivered their unhappy report. GORILLA NO LADY; WEDDING IS OFF bannered the Philadelphia *Daily News.* The *Record,* not to be outdone, concluded its article: "All wedding gifts will be returned."

Back in the Lintzes' Brooklyn jungle, Massa's old cagemate, Buddy, was getting along no better. In a cruel twist of fate, he was once again the victim of a discontented employee's attempts to get back at the gorilla's owner. This time, a young groundskeeper whom the Lintzes had fired sneaked back onto the property after everyone was asleep and fed Buddy a bottle of a powerfully acidic disinfectant sweetened with chocolate syrup. Buddy innocently quaffed it down. When the Lintzes found him in the morning, he was rolling in agony on the floor. The acid had burned out the lining of his stomach and intestines.

Once again Mrs. Lintz nursed him back to health. He recovered miraculously, but he lost some of his even temperament in the process. One night during a heavy thunderstorm, Mrs. Lintz woke up to see a large dark shape looming over her bed and whimpering. It was Buddy. He had slipped out of his cage and come to his Missy for comfort. Mrs. Lintz nearly fainted at the sight, but she recovered her wits sufficiently to take Buddy's hand and guide him back down the stairs and into his cage.

Now when Ringling Brothers renewed their offers to take Buddy off her hands, she accepted. She specified that Buddy be caged in a twenty-five-foot-long air-conditioned trailer, the sort that rock stars might use for road shows today, and required that they take on her assistant Dick Kroener, to whom Buddy had grown terribly attached, as his keeper.

Rechristened Gargantua the Great, Buddy was featured in extravagant posters as a huge and fearsome killer ten times the size of a man. He was invariably shown snarling—which was all the animal could do because of the acid dousing. In one poster he brandished an African native in one fist like a teaspoon. The legend read:

RINGLING BROS. AND BARNUM & BAILEY
COMBINED SHOWS
THE LARGEST GORILLA EVER
EXHIBITED—
THE WORLD'S MOST TERRIFYING
LIVING
CREATURE!
GARGANTUA THE GREAT

Although he proved a poor match for Bamboo, Massa remained welcome at the zoo, and he remained in the cage next door, from which he delighted in tormenting his frustrated suitor by occasionally showering him with water from his drinking pan.

Quite likely his antisocial behavior stemmed from the fact that he had never before been allowed to be a gorilla. Obviously, one cannot expect that Massa should have welcomed Bamboo's solicitations. Male gorillas in the wild generally have little to do with fellow males, regarding them as unwanted competition for troop dominance. But Massa wouldn't have anything to do with females either. Even after the zoo staff had discovered the secret of his sex, they never thought of Massa as a breeder. He lived out his days in solitary confinement, looking to humans for companionship and feeling estranged from his own kind.

He lived for visits from his Missy. "When I go to see Massa," writes Mrs. Lintz, "she is delighted, beats her chest, and tries to kiss me through the bars. When I leave, she climbs up the front of the cage, throwing kisses until I am out of sight. Whatever our difficult moments, we are closely bound together."

Gargantua died at nineteen in 1949, but Massa lived on and on. Since the circus was interested in producing an imposing spectacle, Gargantua's weight was allowed to balloon past 500 pounds; Massa never broke a trim 375. And Massa was fed Zoo Cake, the highly nutritious, if not very tasty, mixture that the Philadelphia Zoo itself had developed back in the 1930s. Also contributing to Massa's impressive longevity, early on the zoo had installed glass partitions to wall its great apes off from human visitors and protect them from respiratory ailments. Tuberculosis is particularly dangerous, but even a common cold can bring misery to an entire troop of great apes.

Bamboo died of a heart attack in 1961, and Massa assumed Bamboo's title of world's oldest gorilla a few years later, extending the record year by year until his death. Massa went into a marked decline in the last years of his life. His weight was down to 175, and the skin seemed to sag off his bones. He looked like an old man. His hair, once shining silver across his shoulders and back, had gone a dull gray everywhere, and it was thinning in patches about his body where he had plucked at it out of nervousness. His joints creaked when he walked. Only three teeth remained in his mouth, the rest having been pulled by the zoo veterinarian in two marathon operations. And his nasal passages had been widened to improve their drainage after he came down with a severe case of sinusitis. After the surgery, one zoo researcher watched him as he fought off the anesthesia. "He would occasionally rise to his feet," he wrote, "stagger about his cage briefly and then collapse in the straw that padded his cage floor. Several times when he arose he appeared to be wrestling with demons and flailed out angrily at those imagined enemies." Afterward, Massa seemed quieter and more contemplative. He would sit in his cage with his chin resting on his right forearm, and his left arm draped over the top of his head. He would gaze deep into the eyes of his visitors. Sometimes, when he felt overburdened, he would turn his back on the crowds that came to see him, and not turn around until they had dispersed.

Then came his fifty-fourth birthday, and the crowds were gone for good.

When word of Massa's death went out across the country and around the world, the zoo received several hundred letters of condolence from individuals, ranging from the keepers of the gorilla house in Tokyo's Veno Zoological Gardens to the nine-year-old in Philadelphia who sent a tiny silk flower in a harmonica box. The staffers also handled a number of inquiries about where Massa's funeral service was to be held and where he would be buried. Responding personally to each letter, Debbie Derrickson and her assistant merely thanked the letter writers for their sympathy. They did not say that, being a gorilla, Massa would receive no service and no burial. Instead, his body would be shipped out to various medical research laboratories around the country. Massa's eyes stayed at the Scheie Eye Institute in Philadelphia, but his brain was divided among Johns Hopkins in Baltimore, Montefiore Hospital in the Bronx, and the University of Pennsylvania; his teeth went to San Diego, his heart to Baltimore, Philadelphia, and East Germany; and the rest of the body would be kept frozen in Virginia until it could be reconstructed by an anthropologist at the Smithsonian Institution. All these institutes and hospitals had made reservations long in advance to get a close look at an animal that was so similar to man, and that had lived to the equivalent of nearly a hundred human years in a controlled environment. "Studying Massa is like studying Methuselah," said Dr. Daniel Cowan, one pathologist who would get to examine the body. From zoo records it was possible, for instance, to determine everything that Massa had eaten for the forty-nine years that he had spent at the Philadelphia Zoo—including the occasional Tastykake that Ralph McCarthy gave him.

When I arrived in Philadelphia in July 1985, a sculpture of Massa by Eric Berg stood surround by piles of brick, stacks of lumber, and other materials waiting to be assembled into the new World of Primates exhibit, a splendid open-air habitat that would go up on the very site of the increasingly dreary Monkey House where Massa had lived for so long.

It was odd to see Massa there in bronze, staring off into space, oblivious to all the junk around him. But little by little, as the months went by, the piles diminished, the new exhibit took shape, and Massa's form emerged from the rubble. It was an omen. Like some patron saint, Massa was an inspiration to the zoo staff. His body may have been scattered about the country, but his spirit lived on at the Philadelphia Zoo.

Summer

The oldest zoo in America stands on a bank of the Schuylkill River, ten minutes by car from the flamboyant City Hall at the center of downtown Philadelphia. Some of the Philadelphia Zoo's first visitors on opening day, July 1, 1874, arrived by ferryboat; others came on horseback, by railroad, or by coach. This stretch of the Schuylkill appears in Thomas Eakin's well-known painting of 1871, *Max Schmitt in a Single Scull*. His friend Eadweard Muybridge photographed a number of zoo animals, including a camel, a parrot, a sloth, and a lion, in the 1880s for his studies of locomotion. (Since Muybridge didn't venture into the lion's cage for the photographs—"that was not to be thought of"—the lion was photographed through the bars and ended up, Muybridge noted, looking more like a zebra.) Oarsmen of the so-called Schuylkill navy still ply the waters today. Their boathouses—Vespers, the Philadelphia Rowing Club, and others—are barely visible across the water from the main gate. Outlined in lights just downstream, they give the far riverbank at nighttime a festive, Parisian air. A short way farther down, the Philadelphia Museum of Art, built in imitation of a Greek temple, commands the river for art. It stands over the old waterworks, which once contained a modest aquarium that upon its demise furnished the zoo with a snapping turtle.

Just upstream of the zoo, the river passes through Fairmount Park, at three thousand acres the largest municipal park in the world. The land was acquired in the 1850s to secure safe drinking water for the city. Fairmount Park was the site of the famous Centennial Exposition of 1876, which drew ten million people, fully a quarter of the population of the United States. To the zoo's good fortune, many of them ventured across the street to check out Philadelphia's animal collection, boosting the zoo's attendance to a point it didn't reach again for nearly a century. A few of the exposition buildings—Horticultural Hall, Centennial Hall—still stand. They suggest something of the neoclassical grandeur of the event, but they have fallen into sad disrepair.

The zoo is built on a sliver of Fairmount Park, and it pays the park a dollar a year for the privilege. It looks, at first, as though an amusement park has perched on a corner of the property. A monorail runs around the zoo, twenty feet in the air. Flags and balloons toss in the wind. Children stampede about the grounds, and their delighted squeals mingle with the barks and roars of the animals. Peacocks stride freely across the lawn, resplendent in their shimmering feathers. And the gay colors and outlandish shapes of the many high-Victorian buildings have a fantastical, storybook quality.

The zoo's land is shaped like the side view of a Galápagos tortoise. But the shell, in the zoo's case, offers more confinement than protection. Its crescent shape is defined by the arc of the railroad line that skirts its western border. In the nineteenth century, some European travelers took the zoo's bison herd as evidence that the American prairie extended to the edge of New York City. Originally, the Reading and Pennsylvania railroads, of Monopoly fame, ran by; but now Conrail freight trains and Amtrak passenger liners whistle past on their way to Washington and New York. It is the most heavily traveled stretch of train track in the Northeast. The zoo's impressively dark soil, apparently rich in nutrients, is actually thick with sooty—and chemically inert—fly ash from the original steam locomotives.

The flat underbelly is formed by Thirty-fourth Street and, just beyond it, the Schuylkill Expressway, which runs along the riverbank. The expressway proved to be the final resting place of one of the large and graceful crowned cranes—gray, with a plume of yellow—from the zoo's collection. Despite a pinioned wing, it managed to get a running start and obtain just enough loft to clear the zoo's fence. The trajectory carried it over Thirty-fourth Street, but dropped it on the highway. Keepers rushed to the scene, but in seconds the bird was gone, pounded so hard into the asphalt nothing remained except a faint reddish stain and a few feathers.

The neighborhood to the west, a run-down section of the city called Mantua, is nearly as hazardous to the animals' health. In one case, a gang of toughs from the neighborhood broke in to stone the flamingos, snapping the fragile legs of several of them. Another time, some vandals spray-painted the tortoises. And the zoo has regularly had to contend with petty crimes against its patrons—purse snatchings, car break-ins, and other hazards of urban life—that originate with their Mantua neighbors.

At forty-two acres, the zoo is only slightly smaller than the fifty-five acre average of the 175 zoos in the American Association of Zoological Parks and Aquariums. But most of the zoos in the AAZPA are small municipal parks with which the Philadelphia would not in any way like to be compared. By most analyses, it stands high in the second echelon of zoos, the parks at St. Louis, Milwaukee, Cincinnati, Chicago, and Seattle, based on the number and variety of animals, the quality of their care, the condition of the zoo's facilities, and the general fame of the organization. But it dreams of the big time, of becoming a national zoo, and that means attaining the level of the Big Three: the San Diego Zoo, the Bronx Zoo, and the National Zoo in Washington, D.C. All those places are monsters compared to the Philadelphia. San Diego has 100 acres in Balboa Park, and a sprawling 1,800-acre Wild Animal Park in San Pasqual that re-creates a good deal of Kenya on the northern fringe of San Diego. Its herd of white rhinos is the second biggest outside Africa. "We make pilgrimages to San Diego," says Philadelphia Zoo director Bill Donaldson. The San Diego Zoo's outlay for public relations equals the Philadelphia's entire operating budget. The Bronx Zoo is set on 265 acres. Philadelphia could fit its entire park inside the Bronx Zoo's new Wild Asia exhibit. The Bronx also runs the New York Aquarium, a laboratory of marine sciences, the Central Park Zoo and two other small city zoos, a worldwide conservation program called Wildlife Conservation International, as well as a research facility at St. Catherines Island off Georgia. The National Zoo is part of the 1,700-acre Rock Creek Park, and owns 3,000 more acres for a breeding farm and research center in Front Royal, Virginia. The National's entire budget is paid by the federal government. The bulk of Philadelphia's budget is paid for by gate receipts. And the National Zoo receives other perks as well, such as the two famous giant pandas Ling-Ling and Hsing-Hsing that were donated by the Chinese government and that some officials of the Philadelphia Zoo have taken to calling, enviously, the "giant marshmallows."

By contrast, the Philadelphia Zoo is a tight, cramped shoebox of a place with barely enough room to turn around in. Larger zoos can breezily add more exhibits, and give more space to the ones they already have. For the Philadelphia to expand, another part of it has to contract. A large map of the zoo adorns a wall in most of the administrators' offices, and the various vice-presidents fight over space like opposing generals. If new construction is to go up, a wrecking crew has to come in first. Often it discovers the remains of yet a previous building on the site. Because of the rapid pace of change in animal management, zoo buildings go out of date long before they wear out, and they tend to wear out rather quickly. The zoo's Rare Mammal House was up-to-date when it was constructed in 1965, but with its nearly pathological interest in sanitation, it is laughably quaint today. The floor is made of bathroom tiles; and the plumbing enables the cages literally to flush.

This summer of 1985, behind a sign reading PRIMATES AT WORK that the contractors did not appreciate, the Rare Mammal House was being replaced with a new state-of-the-art exhibit, the World of Primates. The product of three years of planning and over $6 million in contributions and loans, it is an open-air exhibit that will allow various primate groups like gorillas and orangutans to romp together on several moated islands in an artificial lake. If the trend in the sixties was toward sanitation, the trend in the eighties is toward open, naturalistic exhibits. The World of Primates takes the old-fashioned cage and turns it inside out. Instead of tiles, there is grass. Instead of walls, there are watery moats. Instead of bars, there is air. The human visitors are the ones who'll be hemmed in—they'll watch the animals from narrow viewing stations around the periphery of the exhibit.

As I watched in July 1985, the giant yellow Caterpillar tractors groaned and strained as they clambered about the site, shoving the earth around and sculpting the zoo into still another shape. Trees were stacked sideways like cordwood as they waited to be planted to turn the landscape green. Workmen moved about in workboots and hard hats hauling wire and laying bricks for the accompanying animal house. Among them, scrambling around in his gray suit and dress shoes, was the business manager, Rick Biddle, watching like a hawk to make sure the zoo got its money's worth.

The zoo is an organism, too. Right now, one of its cells was replicating.

That summer the zoo had 1,700 animals from 550 species in its collection, more or less. No one knew exactly how many animals there were. The zoo had a pretty good idea about the larger mammals, of course. It would be hard to keep the birth of a polar bear or a hippopotamus, for instance, a secret for very long. But some of the pouched mammals, like the kangaroos and wallabies, might give birth and quietly harbor their tiny young for days without anyone noticing. The reptiles and amphibians, placed in solitary confinement for the most part in glass cages, couldn't procreate so freely. But many of the birds were at liberty, either at the free-flight Hummingbird House, or in large groups in outdoor pens, or nesting along the banks of Bird Lake, and were hard to tally. Every month, the various animal departments swept through the zoo to conduct an official census. And then within days, the count would be out of date again. By general reckoning, though, the population was expanding—births outnumbered deaths by about three to two. Because space was limited, any surplus animals were shipped out to other zoos.

In maintaining its own ecosystem on the banks of the Schuylkill, the zoo declared war on many of the organisms lurking outside its gates. No zoo is an island. As much trouble as it sometimes is to keep the exotic collection in, it is even harder to keep the feral population out. Ironically, both groups sometimes appear equally eager to relocate. Pigeons are the worst offenders. Until the zoo started to take decisive action in the seventies, nearly two thousand pigeons called the Philadelphia Zoo home, and the zoo was losing about $30,000 a year in birdseed. But the zoo also worries about rats, mice, cockroaches, raccoons, opossums, sparrows, swallows, and stray dogs and cats. There was also a certain gray-horned owl that had been seen snapping the heads of the waterfowl.

The zoo employs two full-time exterminators, one of them an unusual-looking fellow who is barely five feet tall and nearly round, to repulse the invasion. It is a delicate matter, of course, to control the pest without eradicating the exhibit animals. There isn't very much they can do to poison the cockroaches in the Bird House, for example, without poisoning the resident birds as well, since the birds eat cockroaches. And the pyrethrum Sectrol can't be sprayed in the Reptile House to kill cockroaches because it kills the snakes, too. Currently, the zoo was using an anticoagulant called warfarin to control the mice. The secret to success was to put it into bits of Zoo Cake, the staple food that the rodents were used to eating, spread around outside the cages. That way the rodents wouldn't get suspicious. For the pigeons, the zoo used Avitrol, a hallucinogen. "It goofs 'em up," said one official, "and before you know it you see 'em walking on the freeway." Strays were controlled by Havahart traps posted at likely entry points around the zoo.

The limiting of such unwanted animals is more than a matter of aesthetics, for some of them are known to carry diseases that can destroy a zoo population. There were alarming reports, as summer began, of the spread of rabies up the Atlantic Coast. One rabid squirrel that sneaks under the fence can bring down any number of zoo animals. An outbreak of rabies had closed the Jerusalem Zoo in Israel for a time in the seventies. And, closer to home, a rabid raccoon was once found inside the panda exhibit at the National in Washington. Any animals at the Philadelphia Zoo that are suspected of contamination, whether they are the zoo's or strays, are killed and decapitated, and the heads sent to the U.S. Department of Agriculture's laboratory in Ames, Iowa, to confirm the diagnosis.

A few other animals find their way in by way of the commissary, but only one has stayed on at the zoo very long, a small pig that was intended to provide dinner for the zoo's twenty-foot-long reticulated python. The "retic," as it is familiarly called, normally squeezes its prey to death with little ado, but it befriended this pig. The snake let the pig live on for days, then weeks. The keepers kept taking the pig out of the snake's cage, then putting it back in a few days later, but the snake wouldn't kill it. Finally, in some mystification, the curator of reptiles granted the pig a reprieve and put him on exhibit in the Children's Zoo.

And the zoo attracts a fair number of the species *Homo sapiens* within its gates. Nationally, zoos draw over 110 million visitors every year, a number exceeding the figure for all American professional team-sports events combined. The Philadelphia Zoo brings in a decent share. For fiscal year 1985, 1,235,000 came through the turnstiles into the zoo. According to statistics, the zoo clientele ranges through all socioeconomic groups, all ages, all colors, and both sexes, but families with small children make up the bulk of the visitors. Surprisingly, the visitors don't come expressly to see the animals. According to the zoo's questionnaires, 70 percent merely want to go on a pleasant outing with family and friends. The animals apparently are only an added attraction. Consequently, the major competition for customers comes not from other nature-oriented activities, but from local shopping malls. Still, the patrons do develop deep affection for the animals they visit. Some of the keepers dismiss the more sentimental ones as "bunny-huggers," "bambiologists," and "humaniacs" for viewing all animals as humans in animal dress.

Anthropomorphism is a dreaded word at the zoo, yet the animals freely engage in the equivalent, zoomorphism: They view the keepers as fellow animals. Animal babies of many species are often imprinted on their human surrogate parents, and herd animals frequently attempt to incorporate their keepers into the herd structure.

On the human side of the question, the temptation to see animals as fellow humans can be irresistible. Massa was a great favorite, but a generation of Philadelphians has grown up with George, the massive Siberian tiger, John, the imposing lowland gorilla, and countless other animals in the zoo's collection. These animals become celebrities of a kind, like baseball players or politicians, and some of the relationships can go surprisingly deep. There is one woman who brings her own chair to sit by the gorilla exhibit for hours on end; another woman, dubbed the Wolf Lady, has observed the wolves almost daily for the last ten years. And the zoo has gotten used to psychiatrists inquiring if a certain animal has died: Their patients often grow attached to one animal and can sink into deep depressions if anything happens to it.

The zoo gets a lot of strange calls, and has come to serve as a clearinghouse on information about animals, often of the most basic kind. One caller wanted to know why his goldfish were swimming upside down at the top of the tank; another was afraid that she had given birth to a monkey because her infant child was clinging unnaturally to her leg; a third called because his Capuchin monkey had started attacking his wife every time she had her period; and a fourth wanted to know what the "strange animal" was outside his house with the face of a pig, the body of a porcupine, the tail of a rat, and webbed feet. The strange animal was most likely an opossum.

. . . And so the great tortoise lumbered on into the future.

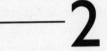

2

Most days, director William Donaldson starts the morning by strolling around the zoo. "I'm a wanderer arounder," he says. He got into the habit in his previous career as a city manager for the metropolises of Cincinnati, Ohio, Scottsdale, Arizona, and Tacoma, Washington, where he went so far as to ride with the trash collectors and the vice squad to see how things were going. But the practice melded nicely with this new job of running the Philadelphia Zoological Society. For the morning rounds are part of a grand tradition in zookeeping, beginning with the great German directors of the nineteenth century and carrying over into the more European of the American zoos, like the Philadelphia.

Donaldson cut an unusual figure as we set out from the granite administration building one bright morning at the end of July. He is bulky. Growing up, Donaldson liked to be called Van, short for his middle name of Vanzandt, but because of his size everyone called him "Moving Van." Today, he wore a Panama hat to keep the sun off his pale skin—"The curse of the Northern races," he said—and a neatly trimmed white beard. He also smoked a sweet-smelling pipe. Because of the white hair and all the insulation, some staffers think of Donaldson as a polar bear. Feature writers have compared him to Santa

Claus, Colonel Sanders, and the Pillsbury Doughboy. He endures such analogies good-naturedly. "It's unfortunate that truth is a legal defense for libel," he says.

Born in the small town of Louviers, Colorado, Donaldson has retained his sunny midwestern disposition in a city known for its chilly social temperament. "When I came here, I felt I was coming *dangerously* far to the east," he admits.

It's oddly appropriate to have a city manager run a zoo, for the zoo is in many ways like a city: It has its own police force (a security detail consisting mostly of retired cops based over the gatehouse), transportation system (the monorail running around the zoo), power station, education department, real estate squabbles, hospital (veterinary), construction projects, budget fights, sanitation crew, restaurants and commissary, public relations issues, political infighting, and resident population (the 1,700 or so animals).

And it has its oddballs. Donaldson was late getting started this morning because of one of them. He had been in a radio studio making an appearance with a woman named Phyliss, who billed herself as an animal psychic capable of "reading" the minds of the animals, and a former National League umpire who came on with two dead goldfish. As a warm-up, Phyliss had made the rounds of the zoo yesterday and distinguished herself by declaring that the zoo's female Indian rhinoceros, Xavira, was barren. Debbie Derrickson, who as PR director had the pleasure of escorting such dignitaries around the zoo, hadn't had the heart to tell her that Xavira was actually twelve months pregnant. For the show, Donaldson had been tempted to bring his own pet albino corn snake, Cuthbert, but, not wishing to give the listening audience the idea that one could keep dangerous snakes at home, he had brought a porcupine called Quilliam from the Children's Zoo. "I figured that would screw her up—a dorky porcupine," he said. He had assumed that Phyliss would seize on the similarities between him and Quilliam—the quills, the pudginess, the slowness of movement—but she had merely declared that Quilliam liked Bill and missed his family. "We'll do anything to get on the radio," Donaldson explained.

We passed out the door of the administration building and onto the wide walkway that circumambulates the garden.

Donaldson is a talker. The first time he ever went on television, he appeared with a couple of engineers to explain a minor bit of engineering to the city of Denver. Cats caught the tongues of the two engineers, and they fell silent after a couple of syllables, leaving Donaldson to carry the show for the remaining half hour. He experienced no trouble whatsoever.

"I ever tell you about Maggie the duck?" he asked now. As usual, he didn't wait for an answer. "Maggie was a wood duck, and she was protected by the wetlands laws. This woman out in the suburbs took a liking to her, managed to catch her, and brought her home. She took all the carpets up and let Maggie run around the house loose. This woman was huge, out to here"—Donaldson put his arms straight out, and puffed out his cheeks—"a wonderful specimen of fatness. But the neighbors complained, so the wildlife commission came down on her and took her duck away. The lady raised a huge ruckus. I mean, she called everybody! The governor finally had enough of it. He said, 'Give her back her fucking duck.' Possibly those were not his exact words. The wildlife people insisted she couldn't have the duck without a permit, and that would take a while to furnish. So they said, 'Leave it at the zoo. When you get your permit, you can have Maggie back.' So we took her duck and put her down there in the Children's Zoo.

"But my God, just after Maggie came, these two appear, a big fat lady and her daughter. And they come with a dresser full of clothes for Maggie. They came first thing in the morning, every morning, and they were interfering with the operation down there. So we finally had to tell them that the duck was in quarantine and they couldn't visit. Next thing I knew they were up in my office, complaining. I was real glad when the state came and finally took Maggie away."

We resumed our journey and passed a well-dressed young black man striding up the walkway. "Hey, Vance! Howya doin'?" Donaldson exclaimed, his standard greeting. It was Vance Washington, who worked in the cash control department. "Just fine, Bill," Washington replied. Everyone at the zoo first-names Donaldson; he claims that to him, "Mr. Donaldson" means his father. As Washington passed, Donaldson explained that Vance's father, George Washington, was a member of the zoo's cleanup crew. And his brother, Larry, was a keeper in the Bird House.

The Washington family is not unusual at the zoo, which is remarkably inbred. Various familial lines enter the zoo, intermingle, and never come out again. Ann Novak, who runs the adopt-an-animal program after spending several years in the Children's Zoo, is married to John Novak, one of the two exterminators. Bill Maloney, the head of animal services, married a cashier. *Her* mother worked at the refreshment stand. Maloney's son, Steven, is in maintenance. Norman Hess, Maloney's predecessor, paired off with a veterinary assistant, Ann, who now is the zoo's surrogate mother. Childless, she has raised everything from kangaroos to orangutans. Bob Callahan, who runs the Children's Zoo, married the gorilla keeper Roseann Giambro. A lion keeper, Chuck Sturtz, married a cashier; another cashier selected zoo driver Roscoe Stotten.

If the staffers weren't married, they were often siblings. The elephant house is run by the McNellis brothers, a pair who seem to challenge each other to set new standards for grizzled orneriness. Then there are John and George Myers. John, at the Children's Zoo, is about as talkative a fellow as you would ever want to meet. George, in Bear Country, can go a whole year without opening his mouth except to eat. In a rare utterance, he termed his brother "the fifty-cent-word man."

Donaldson meandered up the walkway toward the north gate, and motioned toward the camel Bart, who was strolling in a typically ungainly fashion about a triangular paddock up to our right. With his mop top, Bart looked oddly like an aging Beatle. "We got Bart from a zoo in Syracuse," Donaldson told me. "When he came here, they said he was a trained camel, but the first thing he did was kick his keeper into the ditch. I told the keeper he should get a stick from the elephant house and have a little discussion with Bart about what life is like." The keeper preferred gentler methods of persuasion.

He took a few more steps toward the main entrance at the north gate. It was a pleasant, boisterous spot where visitors poured through the turnstiles inside the gable-roofed, stone-sided gatehouse, designed in the 1870s by the prominent Philadelphia firm of Furness and Hewitt. The firm did a number of buildings at the zoo—all of them wonderfully gaudy Victorian extravaganzas with steep roofs and tall, pointed gables that give the zoo its ambience of high-toned gaiety.

Busloads of kids poured through the turnstiles like balls in a pinball machine, and then bounced noisily from one exhibit to another. They stayed just long enough to shriek with horror (at scary-looking reptiles) or delight (at cuddly mammals), then try out their own barks, roars, howls, and moos in an attempt to speak the animals' language. Desmond Morris, former director of the London Zoo as well as the author of *The Naked Ape,* believes that small children prefer large animals; adults like small ones. The converse may also be true. For the big cats, at least, generally eye all the visitors with the same lazy disinterest, except for the smallest, bite-sized children, whom they view hungrily as lunch. As we approached, some kids were foolishly poking their fingers through the wire mesh and into the gibbon exhibit underneath the monorail stop. The gibbons, sneaky little creatures that look like skinny Eskimos with their white-mittened hands and fur-fringed faces, swung blithely around inside the cage and ignored the kids, who yanked back their hands as Donaldson passed by.

We came to a kea perched inside a lovely dome-shaped wrought-iron cage of Victorian style. A ray of sunlight cut through the trees to shine on the dark, olive-colored bird, bringing out the turquoise in its wing tips. The sight stopped Donaldson. "Isn't that beautiful?" he asked. "They nearly wiped out keas in New Zealand because they thought keas killed sheep. Actually, they just feed on carrion. The sheep died from other causes. But isn't the cage something? We used the wrought iron from the old Monkey House, where Massa was for all those years, but the cages don't come cheap. It cost me forty thousand dollars for three of them, but I'd sure like some more."

One of Donaldson's earliest memories is of his father stopping the family Chevrolet to pick up a turtle that was crossing the road. He handed the animal to young Van in the backseat. Somehow, the turtle got loose in the car and ended up on his father's sleeve, nearly causing him to drive off the road. As a boy, Donaldson always had his eye out for animals, particularly reptiles and amphibians, no matter how much trouble they might cause him. By the time he was ten, he had a menagerie of about three hundred animals, including two alligators. "I had a teacher in elementary school named Ella Frey," he says. "She was a very large, untidy old lady, and her classroom was filled with weird stuff. Beehives, wasps' nests, stuffed animals. She made us look at everything and really see it. She'd ask questions about it, and you had to be able to answer them. She changed my whole life. That's when I started collecting in earnest—insects, rodents, snakes, anything I could catch. My parents really suffered. My father laid down the law. 'No poisonous snakes,' he said. Actually, I did have two copperheads in a box under my bed. I had a couple of possums I named Florence and Mabel after my mother and her best friend. I had a crow that was semi-tame. He'd sit on our clothesline and pull out all the clothespins.

"My best scheme was for catching a turkey vulture. There's not a lot of literature on the catching of turkey vultures, so I had to think. I got an old, rotten red snake and looped a noose around him and I took the end of the rope and went and hid behind a bush. When the turkey vulture came down, I yanked on the rope and got him by the foot. What I didn't know was that vultures have very nervous stomachs and they throw up when they get scared. This particular turkey vulture had just eaten a rabbit, and he puked it up all over me, a big green slimy thing. I stank for weeks, but I had my vulture. I kept him in the garage and he hopped around in there. I clipped his primary flight wings so he couldn't get away. I was so proud."

Animals were nice, but he didn't see how he could make a living from them. For a time, he considered becoming an Episcopal priest, and spent two years studying theology at Trinity College, University of Toronto. He served as pastor at a Toronto prison, and that dissuaded him from pursuing the calling. "It was too depressing," he says. He tried courses in clinical psychology. He even worked as the night attendant at a mortuary. His girlfriend played the organ there. He notes proudly that he worked his way up to apprentice embalmer. He studied Gray's *Anatomy* to see how to fit all the insides back in, then he used liberal quantities of hardening compound to make sure they stayed put, and sewed the bodies back up with a softball stitch. He also worked at a trainyard and in an explosives plant, where he loaded boxes of dynamite onto boxcars. "That always made me nervous," he says, "because we had to throw them in. But you know, actually, dynamite is one of the most stable substances there are."

He got started in city government in 1955 at twenty-four when, confused about his future, he took a job in Denver with the city water department. He had to shut off the water of the people who didn't pay their bills. "Our motto was 'One good turn deserves another,' " he says. He became interested in public management and got work in Beverly Hills as administrative assistant in the city's personnel department, then he became the city administrator for Montclair, California. After five years he went to Scottsdale as city manager; five years after that, Tacoma called; four years after that, Cincinnati. He built a reputation for cost-cutting, innovation, and openness. At Cincinnati, he points out, he didn't just *open* the doors to his office, he removed them entirely.

All the while, Donaldson was quietly feeding his avocation. At Scottsdale, he was able to develop a private menagerie of fourteen cats, six turtles, six quails, six owls, six snakes, and twelve dogs. His wife, Ann, a quiet, gentle lady who is as slender as her husband is large, shares his animal devotions. When a burrowing owl appeared on the doorstep with a broken leg, Ann nursed it back to health by bathing it in Epsom salts every two hours around the clock for two days.

Donaldson says that the main reason he went to interview for the Tacoma job was that the city's Point Defiance Zoo had the country's best collection of sea otters. When he went to Cincinnati, he did so on the condition that he be able to spend one day a week at the zoo there. He became close friends with its director, Ed Maruska, once Maruska was convinced that Donaldson was sincerely interested in the animals. When things got fierce in city government, Donaldson could usually be found at the zoo with Maruska cleaning the salamander cages. His official portrait shows him being kissed by an aardvark. It was Maruska who put Donaldson up for the Philadelphia job.

We traveled on in silence a little while longer, skirting the broad lawn by the zoo entrance, taking a turn by a Victorian gazebo, then swinging around by the granite statue of an elephant with her calf, which, Donaldson informed me, weighed even more than elephants themselves, thirty-seven tons. It is one of several dozen sculptures of various animals scattered around the garden.

Then we took a turn inside the Rare Mammal House. Donaldson made goo-goo eyes at an adorable little chimp, then turned away and said under his breath, "If that little bitch had half a chance, she'd bite my hand off." He recounted the story of the director of the Washington Park Zoo in Portland, Oregon, who used to amuse one favorite chimpanzee by putting his finger in the chimp's mouth to suck on. Sure enough, one day the director stuck his finger in, and this time the chimpanzee bit it right off. Donaldson sent him a plastic finger on a vibrator so he could continue the practice more safely. "He didn't think that was funny," said Donaldson with a chuckle.

We moved on down to the gorilla end, where the massive silverback John was standing guard over his family. Donaldson admired John's physique for a moment, then told me about the sad plight of Bushman, the big male at Chicago's Lincoln Park Zoo who was so fat—over six hundred pounds—he was effectively impotent, since his penis didn't reach past the rolls of fat. To judge by John's two rollicking, dewy-eyed babies, Chaka and Anaka, John didn't have any such problem.

"We ran a newspaper contest to come up with the name of Chaka," Donaldson said. "Well, we didn't know it, but it turns out that Chaka is the name of a Zulu warrior king. When word leaked out about that, somebody threatened to blow up my car. They were all bothered because it looked like we had named a gorilla after a black. It's this old thing about blacks being like gorillas, and it's ridiculous. Blacks are the most evolved of all the human races. They have far less body hair than Caucasians. And who has ever seen a gorilla with thick lips? Anyway, this woman called up screaming at me. I told her, I wish somebody would name a gorilla after *me*. She wanted to talk the whole thing out, so I told her to come on over. When she got here, I took one look at her and I couldn't believe it. She had on an ivory bracelet and alligator shoes, and she was carrying a leather purse. I said, 'Who the hell are you to complain about our treatment of anybody?' Then we both started laughing.

"Now, I can see getting angry if somebody named a chimp after you. Chimps get hysterical. They're like people. They

scream and throw crap. But gorillas? Look at them. They are
probably the most pleasant animals in the zoo. They're calm.
They're vegetarians. When the females go into estrus, everybody mates with them and it's all perfectly fine. There's no
unpleasantness. Gorillas are just nice. I never met a gorilla I
didn't like. And I've met a lot of them, at least forty."

We strolled along by the hippo yard, then back inside the
Reptile House, where he took me into a back room to let me
have a look at a prize Japanese giant salamander, mottled black
on gray, in its private tank. Long and round and warty, it looked
like a baseball bat that had been left out in the rain too long.
Donaldson gazed down at it in the water as if he wanted to pet
it. "Isn't he beautiful?" he said. "I just had to show you."

3

Small as the Philadelphia Zoo is, you can easily get lost in it, and in some ways the grounds appear designed expressly for that purpose. The walkways are laid out in the opposite of the grid pattern that governs so many other zoos; they form a tangle of wavy lines. You can't get anywhere fast. This can be infuriating to first-time visitors, who might expect the signs to lead them directly to their destination. But the Philadelphia Zoo isn't like that. Its signs say, in effect, the muntjac yard is somewhere over there, the Reptile House is sorta that way, as if the journey were the important thing, not the arriving. And so the walkways around the zoo meander, poke along, dawdle.

The zoo is meant to be taken in as if the grounds were your private estate. Its layout borrowed heavily from the tradition of the English landscape designers in its formative days. It was laid out around a pond, called Bird Lake, with a neo-classical temple, the Bird House, rising up behind it. Gazing across the water from Tiger Terrace, the zoo's outdoor cafeteria, you might think for a moment you are at one of the grand country estates, like Stour Head in the south of England. The view is wild and luscious, particularly in the spring-time when the cherry trees are in bloom and are bent over, either from the weight of their voluptuous blossoms or in ob-

57

eisance to the ducks that paddle and the swans that glide about the water.

On the northern side of Bird Lake, where Donaldson and I just strolled, the zoo is more tightly constricted, like a city park, Regent's Park perhaps, the home of the London Zoo. The paths are shaded by London plane trees, or sycamores, and trail past spreading lawns and wooden benches in the English manner. To the south, the terrain grows wilder. The African Plains exhibit stretches out, against a rocky backdrop, with zebras and giraffes parading about in front. A nature walk called Penn's Woods is sandwiched in here, a tangle of evergreens that presents Pennsylvania the way it might have looked to William Penn in the seventeenth century. The wolves prowl about the pine trees just beyond.

This zoo is a garden in other words. Zoos are often called animal gardens, but the Philadelphia Zoo is one of the few to take the idea seriously. The wrought-iron sign out front welcomes visitors to the GARDEN OF THE ZOOLOGICAL SOCIETY OF PHILADELPHIA, and the grounds are called "the garden" familiarly by the staff. Most cities keep their botanical gardens separate from the zoos. In New York, for example, the botanical garden is across the street from the Bronx Zoo. In Philadelphia the two are intermingled. Everywhere there are rose beds, azalea bushes, viburnum shrubs, dogwood, apple trees, and hundreds of other flowering plants. Indeed, the horticultural holdings are more valuable than the animals, if only because the replacement costs are so high. And along with the animals, they give the zoo its seasonal colors—the delicate pinks of the flowering fruit trees harbinger the spring; the fluttering green leaves declare summer; the maples blaze a fiery orange to announce the fall; and the evergreens, with only a sprinkle of red holly berries, stand starkly against the snow to proclaim winter.

And as in a garden, everything is artfully arranged to look natural. The truth is, the zoo doesn't draw from nature half what it takes from Walt Disney. Almost none of the rocks are real but are made of Gunite, a cementlike substance that is slathered onto a mesh of chicken wire and then splattered with gray and black and brown paint in the manner of Jackson Pollock. Behind the gorilla yard outside the Rare Mammal House, the Gunite rocks bear, curiously, the imprint of the Pyrenees. The exhibit designer had been vacationing in Spain with his girlfriend when he saw the perfect shape for a gorilla exhibit: definitely rocklike, but completely without handholds. He made a fiberglass impression of a cliff, rolled it up, and carried it back to the zoo.

In what traditionalists must view as an alarming development, the zoo opened a new exhibit over the winter that contained no live animals at all and little genuine vegetation. This is the TreeHouse, an attempt to show children a view of the world from an insect's perspective. It consists of oversized egg-shells, hollowed-out trees, giant blossoms, and huge honey-combs all done in synthetic sculpture. It is located in a specially renovated Furness-Hewitt building, the old Antelope House. The TreeHouse was created almost singlehandedly by Steve Izenour, a senior associate at the post-modernist architectural firm of Venturi, Rauch and Scott Brown. "Essentially this is the black box," said Izenour, whose father was a theater de-signer. "It's a classic trick of illusionism—you step into a big shed and you're off in another world. You don't have to worry about the real world at all." His major difficulty was in deter-mining the level of illusionistic detail, for at some point he had to blur some of the details or the zoo would go broke. Still, the illusion goes remarkably far. Izenour believes it had to. "With all the competition from Sesame Place, Disney World, and the other theme parks," he says, "zoos are going to have to get more illusionistic to be believable."

The rest of the zoo carries out the same idea. For the in-dividual exhibits aren't so much cages as they are mini the-aters, with backdrops and foregrounds and, where there are high, overarching trees, a kind of proscenium arch defining the space. This conception, curiously, transforms the animal under exhibit into a kind of actor: The camels, polar bears, and cheetahs are, in effect, playing the roles of wild animals in urban Philadelphia.

If the cage is a stage, Chuck Rogers is in charge of the scenery. He is the zoo's resident horticulturalist, the first one hired by any American zoo. He's been here since 1948. "They can't fire me," he says. "I know where all the bones are bur-ied." He wears Panama hats, tweed jackets, and an impatient expression. He operates out of an office strewn with flower-pots that adjoins the greenhouse behind the Small Mammal House.

Showing off his handiwork one day, he pointed to the elephant yard, where the elephants paced around in the dust. "Can't plant anything in there," Rogers grumbled. "The elephants would eat it." But beyond the reach of the elephants' trunks, he had sown a fringe of pampas grass to imitate the tall grasses of their native Africa. The elephants had learned to reach well beyond the sharply pointed wall of rock that marked the official limit of their territory. With their trunks, they positioned a bale of hay to cushion their feet on the sharp rocks. Then they stepped up on it and scarfed up practically everything in sight. Rogers had managed to plant some of the pampas grass beyond the fifteen-foot-wide dry moat that stood as the last line of defense against an elephant exodus. Rogers was a little irked right now because somebody—he figured a human—had run off with the grass's purple plumes. From a safe distance, the yard was shaded by locusts that recalled the exquisite, horizontally growing acacias that make Africa look African. The schist façade of the Pennsylvania Dutch barn that formed the backdrop of this little tableau jarred the spell. But that might have been a conscious joke played by Paul Cret, the barn's designer, tweaking people for believing even for an instant that they were in Africa.

But so it went all around the zoo. Rogers had planted more locusts in the giraffe exhibit some years ago. He had put in twenty-footers, and prayed that they would grow faster than the baby giraffes that were then the exhibit's only occupants; the trees just made it. He had also planted some fire thorn, a North American version of African thorn bushes, along the fence in front of their moat. That hadn't done as well. Giraffes are supposed to be frightened of water, but they had been quite willing to get their feet wet for Rogers's fire thorn.

Indoors, the American alligator exhibit was planted with palms and grassy plants to look like the Everglades. The alligator's neighbor, the American crocodile, had Spanish moss and bromeliads to make his pen look like an estuary. Just outside the Reptile House, a paperbark mulberry had been planted in the tortoise yard in a vain attempt to suggest the broad-leaved palo santo trees of the Galápagos Islands, trees that, like the acacia, wouldn't survive the deep-freeze winters of Philadelphia. And he let the weed tree ailanthus grow freely over the tropical animals, since, with its long, fingerlike leaves and epaulet seedpods, "it has a tropical feeling." Besides, it blew in on the wind and took root all by itself. Occasionally, Rogers would add a dab of color to an exhibit simply because it went so well with the animals. With this in mind, he'd dropped in a bed of yellow iris in the flamingo yard. Unfortunately, he couldn't water the iris for fear that he would upset the flamingos' breeding.

In his efforts to landscape the zoo, some of the hoofstock were silently in league with Rogers: They inadvertently planted clover in their exhibits by way of their droppings. But for the most part, he was up against some formidable adversaries. Most of the animals eat just about anything they can get their jaws on, if they don't trample it into the ground first. The tubby, bottle-nosed Malayan tapirs (distant cousins of the rhino), for instance, were fussy eaters, but they had a heavy tread. Wolves had chewed the bark off all the pines in the Wolf Woods, except for the Douglas firs, which they had liked so much they made their dens under them, ripping up the roots.

The peafowl—peacocks and peahens—that wandered freely about the zoo were Rogers's nemesis. There were about thirty of them, and they delighted the visitors, especially when the males spread their magnificent tails and shimmied in a sexual come-on during breeding season. Unfortunately, peafowl love marigolds. Rogers had planted a big batch of them once. They disappeared down the birds' gullets—blossoms, leaves, stems and all—in forty-eight hours. The big birds also had an alarming fondness for petunias, particularly pink and rose-colored ones. The white and purple they left alone for some reason. They also feasted on tulip bulbs and gulped down the heads of the tree peonies. But then they got theirs; for some of the zoo's earthworms, which peafowl also found tasty, carried a disease called the "blind staggers" for the condition it induced in its victims. Rogers may not have been that sorry to see some of the peafowl struck.

—4—

Dan Maloney is one of the new breed of keepers at the zoo. He displays an engineer's dedication to his job, and he handles his animals the way a computer jock might work an IBM PC— with a learned concentration that is difficult to distinguish from love. Just shy of thirty, Dan is good-looking with his shock of dark hair, and so fast-talking he frequently talks himself hoarse. No relation to senior keeper Bill Maloney, whose son Steven works in maintenance, Dan came to Philadelphia from the Bronx Zoo early in the summer. He had thought he would be taking care of the elephants, one of the most challenging jobs at the zoo, but after he arrived, another keeper named Gene Pfeffer expressed interest in the task, so Dan was relegated to Gene's old job with the white rhinos, ostriches, antelope, and giraffes at African Plains.

Despite his disappointment, Dan had taken to the job with a dedication that made heads turn. Not long after he arrived, he asked his supervisor, the assistant senior keeper Dave Wood, for some scouring pads to scrub down the walls of the antelope pen.

"*Scouring pads?*" asked Dave in disbelief.

"Yeah, you know, Brillo pads or something."

Dave found him something, which was lucky. Otherwise, Dan would have licked the place clean.

Although he had only been in Philadelphia a month or two, other keepers had already started to tease him, which was a good sign. "The world's only yuppie zookeeper," the gorilla keeper Bob Berghaier called him. Dan called Berghaier "Bergie." And he called me "Sedgman," a term that other zoo people soon picked up as well.

Dan had left the Bronx because on his salary he couldn't afford to buy a house in the area. In Philadelphia he and his wife, Michelle, bought a town house by Fairmount Park not far from the zoo, and the two of them were spending all their nights and weekends fixing it up. Unlike the traditional zookeeper, Maloney has a college degree. He sometimes feels a need to keep quiet about it. When Gene Pfeffer asked him if he'd been to college, Dan admitted it was true, then added, "But it didn't ruin me." That drew a laugh from Gene. Dan majored in biology at Kutztown State College in Pennsylvania. That's where he'd met Michelle, a graphic artist. Because of her, he traveled in higher circles than the usual zookeeper, and he sometimes found himself at some swank parties. "At the parties, I try to look pretty much like everybody else," he said. "I have a glass of wine in my hand, I'm munching on cheese, and I'm watching my double negatives. I'm viewed as something exceptional, and I like that. Everybody else is an analyst of something—systems analyst, engineering analyst, account analyst. When I say I'm a zookeeper, it sets me apart."

Dan picked up his interest in animals from his father, who worked as an FBI agent specializing in organized crime but who loved to go out into the wild with his son. "We always had creatures around the house," Dan remembered. "Never the usual pets. I always had snakes, turtles, and other stuff my dad and I picked up tromping around swamps. I raised baby squirrels. They were fun. They'd come down out of the trees when I called. I kept one of them in a cage in the house. I'd let him out in the morning when I went to school and when I got home, I'd clap my hands and yell out, 'Hey buddy!' and he'd come in the window and jump up on my shoulder. I probably should have let him stay outside once he was grown, but it was just so cool to have him."

I asked him the squirrel's name.

"I never named him. He was the only one I had in the house, so I didn't feel any need to name him. I don't think there is any need to name any animal, really, except maybe elephants, sea lions, and primates, since they respond to a name.

"I also had a skunk—that's a neat animal," he went on. "And a woodchuck for a while. The thing was, once I got to an age where I could catch these creatures, that's when I realized it was kind of a crime to keep them. It was a paradox. Once I could catch them, I didn't want them anymore. Now I have a couple of flying squirrels and two caimans, you know, crocodilians."

Sedgman expressed surprise.

"Just little ones. They're in my mother's basement now in a box. A friend of mine gave them to me. I call them Jake and Elwood. You know, the Blues Brothers. They are terrible animals. They have a terrible disposition, really aggressive. My mother doesn't like them all that much. She only has them because I don't have room. They have gotten out a few times. When that happens, I usually go get them. My little sister is good with them, too. She's fearless. I always used to take her with me to catch things. She'll always just reach out and grab it. You catch them right under the head. They're strong for their size, but I don't think they could take much of a bite out of you. I don't think they could even take a finger off. They could probably break it though. They have a twisting motion that pulls. I once came home and found my mother and brother and sister all armed with baseball bats. They were all going down to the basement to make an assault on the caimans because they'd gotten loose. So I went down too, and we grabbed them. It was no problem."

Dan took me down to African Plains one afternoon around quitting time, and I watched as he brought the animals in for the night. The animal buildings are concealed behind a Gunite rock façade. Inside, there are some large barred cages with cement walls. The rooms were extraordinarily neat, more like storage facilities than animal pens. Some shovels, rakes, brushes, and brooms lined one wall, all of them plumb vertical. The white rhinos, squat, ungainly creatures with horns that can reach five feet but are usually worn down in zoos, were already in place behind thick bars. "White" is a misnomer—the beasts are actually a rocklike gray.

Dan brought the giraffes into their pen through a doorway that was twenty feet high. Since it was a cool day, the giraffes were grouped by the door waiting to come in. On a warm day, they usually like to stay outside. Giraffes are herd animals, and they do everything in a group, and that can make it hard to move them in and out. None of the giraffes is willing to take that first decisive step. "But if you get a good lead animal, you're all right," said Dan.

"C'mon Twig, c'mon buddy," Dan called out gently, projecting calm with his voice. Twig was short for Twigga. She was his lead animal. Dan clapped his hands quietly as he called to her. Finally, she came stalking in. For all her height, she retained a wonderful elegance, like an extremely tall woman. She was slightly gawky, but the suggestion of frailty in something so big is seductive. The frailty is an illusion: One kick from her legs could drop a lion. Chestnut brown in color, Twigga was considered handsome for a giraffe. The lines of the squiggly checkerboard pattern on her coat were especially slender. Her neck was straight, not humped like some, and her proportions were right. Twigga walked in a kind of dance step that seemed somehow Egyptian—her head bobbed forward, then her chest surged, and finally her front leg swung ahead, all in a sweeping wave motion that accorded with some ancient law of physics. "She's good-looking," said Dan. "But she's kind of a deadhead."

After Twigga came in, as Dan had expected, the others trooped in after her—Puzzles, and Twigga's new month-old baby, Amanda (named for Gene Pfeffer's daughter), already a nimble six-footer.

As Puzzles, the male, spied me, his head dropped down to mine as if lowered from a crane, angling down and down and down. We nearly touched noses through the bars, and I could feel his warm breath on my cheek. Puzzles' head was crowned with two knobby horns, his face adorned with bristly hairs, and he had a few other bumps around his eyes. A giraffe bull continues to lay down layers of facial bone well into maturity. Possibly they help protect him in his ritual head-bumping battles

with other males. (When not bumping heads, male giraffes en-
twine necks with each other in what appears to be a lovers'
embrace, but is actually a dominance struggle akin to human
arm wrestling.) His breath tickled. When I reached up to rub
my cheek, Puzzles panicked. He lofted his head once more up
to the ceiling. "Giraffes are real skittish," said Dan.

Then the ridiculous ostriches, Gus and Gertie, strode into
their enclosure on their thick and powerful legs. They wrapped
their feathers around them like boas. "You got to watch yourself around these guys," said Dan. "They've got legs like Eric
Heiden. They can really kick." He dangled some dead mice in
their direction. Each swallowed their portion in a gulp, then
moved on to a salad course that Dan had prepared for them.
Fidgety creatures, neither of them could relax with their dinner, but kept bobbing their heads up to see if I was still there.
Their flat-billed mouths were hinged nearly at the back of their
heads, like Mr. and Ms. Pac-Man's.

To help me get back into the giraffes' good graces, Dan
handed me a carrot to give Twigga. She grabbed it with her
scratchy black tongue and then skied her head again while she
gulped it down. She and her baby, Amanda, continued to eye
me suspiciously for some time, however, as if they had never
seen anything quite so strange as a human being.

5

It had been a bountiful spring. After Massa died, the zoo inhabitants had been engaged in a reproductive binge unparalleled in recent memory. The birds in the Bird House had been busy laying their eggs, as had the ducks and geese out on Bird Lake, forcing the keepers to paddle out in canoes twice a day to gather up all the eggs before they hatched. Otherwise, the ducklings and goslings—not yet pinioned like their parents— could have taken wing and drastically altered the ecology of the Delaware River Valley.

But the profusion of mammals was the exciting part. Blue Eyes, the four-year-old Malayan tapir, had produced a darling little porker the keepers teasingly called Denise, after the slim keeper Denise Robinson in the Reptile House across the walkway. The birth seemed particularly providential, since Bill Donaldson had been given Spook, her father, on a breeding loan by the Rotterdam Zoo as a way of welcoming him to the zoo business. Over in the Carnivora House, Beauty, a female binturong—a shaggy, black catlike creature, also called a bear cat, of the civet family—had given birth to a pair of young. When they failed to gain weight, suggesting that Beauty was being an inattentive mother, they were removed for hand-raising. That set off a scramble among the human females over at

69

the Penrose Laboratory as to who would get to be the surrogate mom. Eileen Gallagher won out. Because the male seemed
always to be screaming, she named them Ralph and Alice after
the Kramdens of *The Honeymooners.* Alice was quickly shipped
out to the Cincinnati Zoo, where she would be trained to be
the mascot for the University of Cincinnati Bearcats. Ralph
stayed in Philadelphia, and he took to Eileen like paint to a
wall. Eileen didn't exactly keep Ralph; she wore him—on her
shoulder, usually, but sometimes on the top of her head. In a
private exhibit outside the Small Mammal House, a lesser
panda—a ruddy-brown foxlike cousin of the famous giant
panda—had delivered a pair of little panda babies. They were
being raised during the day by Ann Hess in the nursery. Eileen took them home at night.

And most exciting of all, there was another gorilla in the
troop. While gorillas tend to be secretive about most of their
personal affairs, they are remarkably open in their sexual behavior. Chimps, in contrast, are all too open about everything
except sex. None of the keepers, for instance, had ever seen the
chimps in the act of breeding, although they had seen them in
the act of nearly everything else. Roseann Giambro, a short,
forty-year-old with dark hair flecked with gray that was almost
precisely the color of a gorilla's coat, had insinuated herself so
deeply into the affections of the gorillas she could virtually
count herself one of the troop. She had seen the pudgy, arthritic
female, Snickers, mate with big John, and she knew Snickers
was pregnant. Roseann sensed that Snickers had fallen into a
bad humor, but hadn't suspected she was close to giving birth
until a visitor came up and asked her how the little baby was
doing. "I thought she was talking about Chaka, Samantha's
baby," Roseann told me, "and I said he was fine, and she said,
'But he's so little.' So I went over to take a look, and Snickers
was just passing the afterbirth. There was a bit of blood in the
cage, and Snickers was pacing around nervously, but otherwise
everything was fine." Because of the public relations value (never
very far from the minds of the zoo administration), the business of naming such prominent babies is relegated to the "gorilla editor," Frank Dougherty, at the Philadelphia *Daily News,*
who runs a competition. That's how they had come up with
the unfortunate selection of "Chaka." This time, the name

Rambo actually topped the list, but the naming committee set-tled on Anaka as being the most appropriate. Now, two months later, Anaka was fleshing out nicely. Although gorillas gener-ally mature slowly, like humans, Anaka was at two months al-ready fully mobile, and she had most of her teeth.

The biggest and most long-awaited birth was still to come. That was the rhino baby, fathered by the zoo's own Billy and mothered by Xavira, a rhino lady who had been flown in from Zurich for a little romance. Donaldson called her Billy's Swiss Miss. Now that the two had had their fling, they had been separated into private stalls alongside the elephant house, for the long wait. Solitary in the wild, rhinos were supposed to like it that way, but I often saw Billy raise his massive head over the fence to take a peek at the lusty Xavira trotting about her yard. By checking Xavira's ovulation cycle, and only allowing Billy in for limited conjugal visits, the zoo had a pretty good idea that conception had occurred back in mid-February 1984. But the prediction of delivery dates for Indian rhinoceroses is hardly a science. Only eight Indian rhinos had ever been born in captivity in the United States, as opposed to their more common relatives the white rhinos, which had reproduced in-numerable times. The books say it takes 450 days for Indian rhino gestation, which would place the birth in early October, but no one would be surprised if the baby dropped in Septem-ber or November. Even August or December couldn't be ruled out; the Los Angeles Zoo's rhino had been two months late.

6

But the zoo couldn't settle down for the rhino watch just yet. It had another advent to prepare for—the coming of the koala, the first appearance of a koala on the eastern seaboard. This one, an adorable one-year-old named K'bluey, was a loaner from the San Diego, one of only three zoos in the United States (the Los Angeles and San Francisco zoos were the other two) to exhibit the creatures, which by most accounts were second only to giant pandas as the best draws in the business. In the parlance of the trade, koalas were "celebrity animals," turning a magic combination of rarity, cuddliness, and regular appearances in Qantas Airlines commercials into stardom. The San Diego had loaned one, K'bluey's father, Walt (short for Waltzing Matilda), to the Cincinnati Zoo for a month the previous year and the animal had pulled in 100,000 people. Komodo dragons—reptiles from Indonesia that look like small-scale dinosaurs—are also pretty good draws, but they're scary and that tends to limit their popularity. Still, the Philadelphia Zoo was trying desperately to extract one from the Indonesian government for a new display. White tigers are also popular. The Philadelphia had had a couple in from Cincinnati for the summer last year, and they had boosted attendance by 55,000. With the koala, business manager Rick Biddle was banking on even more than that.

In preparation for the visit, the "Koala Kommittee," headed by twenty-five-year-old Steve VanderMeer from the marketing department, had been meeting for months now, and getting a little silly. In its eagerness to maximize the koala's sales potential, critical to a zoo that relied so heavily on gate receipts, the Kommittee had lined up an impressive list of spin-offs and promotions by which other businesses could ride on the cute and cuddly image of the koala, and the zoo could use the advertising to get the word out about K'bluey's arrival. Fotomat, radio station WDAS, the fast-food franchise Popeye's, and the supermarket chain Superfresh, Qantas (naturally), and American Airlines were all getting in on it. In an ambitious step, the zoo had also taken out billboard ads—depicting a koala in silhoutte, the image to be gradually colored in as K'bluey's arrival approached—and television commercials; and even airplanes dragged the message across the sky for vacationers at the New Jersey shore.

It was an aggressive undertaking for a zoo long marked, like Philadelphia itself, by genteel passivity. The project began with the business office, and it began, ironically enough, with a Biddle. At the zoo some people had taken to calling Rick Biddle "The Hatchet," and the term pleased him. With all the zeal of a hard-nosed CEO, he was determined to cut costs and raise income. He crunched numbers, sometimes heads.

The Biddles had been nearly synonymous with Philadelphia's banking establishment ever since the days of Nicholas Biddle, a lawyer and literary man who had served as president of the controversial Second Bank of the United States under President James Monroe. Biddles had been a common sight on the zoo's board of trustees through the years for that very reason. Rick made it clear that he was not one of *those* Biddles. "I suppose we're all related on some level," he admitted. "But back in the seventeenth century there was a split between the Protestants and the Catholics. The Protestants have all the money. I'm Catholic." Donaldson liked to give Biddle a hard time about it. "Rick's a Biddle of Philadelphia," he would say with a knowing look when introducing him. He also teased Biddle about his business degree from Villanova. "He always claims he can't remember the name of it," said Biddle. "He says, Rick went to that school up the highway from here, you know, the one that's good in basketball."

Biddle attracted such affectionate teasing because he comes across so tough, you feel compelled to try and get him to loosen up. He is lean as a hunting dog, and his jaw hangs loose in a permanent scowl. He has the look of someone who has been up late over the company books, and is irritated by what he has seen. He doesn't sleep well. "I think too much," he says. Something of an amateur photographer, he specializes in pictures of sunsets, some of which are displayed on his office wall. Why sunsets? "Because I know what is going to happen next."

Only twenty-nine, Biddle is one of the young bucks who are pulling the zoo into the modern age. If Donaldson is Santa Claus, then Biddle is one of the reindeer hauling the sleigh. He came to the zoo from a federal subcontracting firm in artificial intelligence. The company had grown, he said, from $2 million to $10 million in three years, and he had risen from accountant to comptroller, but he felt stymied. "I didn't agree with the direction the company was going," he said. An employment agency told him there was an opening at the zoo where he might have more room to maneuver. He didn't know a thing about animals when he arrived. He had to ask what the "big pink birds standing on one leg" were out in front of the administration building. He was surprised to learn they were flamingos. After five years at the zoo, plus two trips to Africa with the Zoofari Club, he still didn't know much about animals except how much they cost.

He did know about numbers, though, and right now the numbers worried him a little. Too much was hanging on the koala. Even though the zoo's fiscal year had just started on July 1, Biddle had turned the matter over and over in his mind, and the entire year hinged on one month—August. "It's my theory of momentum," he explained. Simply stated, the theory is that if things start out well, they'll get better; but if they start badly, they'll get worse. The Philadelphia Zoo, Biddle knew, didn't have much resilience to adversity. As a private, non-profit institution, owned and operated by its board of trustees, the zoo did not have the luxury of a municipal government to bail it out during down times. (On the other hand, it didn't have to run to the mayor for its annual appropriation every year either.) Instead, the zoo depended on gate receipts, and on purchases of food and gifts once patrons were inside. This meant the zoo was at the mercy of the weather to an alarming extent. One rainy weekend and the zoo would be out $90,000. The zoo was severely pinched in the spring of 1982 when it rained on fourteen out of seventeen weekends. In an attempt to divine what the heavens had in store for him, Biddle had even taken to consulting the *Farmer's Almanac*. "I just look at it," he said defensively. Biddle studied the attendance charts carefully. The zoo's equivalent of the Dow Jones ticker, they told him how he was doing. He got the previous day's count every morning, and, every month tabulated the results on long pages, correlated with the day's temperature and general weather conditions. He distributed the daily figure to the senior staff and posted it by the employee entrance. "It gets 'em fired up," he said.

For the fiscal year to turn out well, then, it had to start well. In the northern climate, the zoo did most of its business from May to September, although it had recently been doing better in April as well. August would probably do okay; it usually drew 150,000. But to get the little edge that would set up future growth—generating money for advertising, for improvements, for new facilities—Biddle wanted something extra, and that's where K'bluey came in. The other seasons were taken care of. He didn't expect much from the fall, and the winter would be boosted by the new TreeHouse. And next June the zoo was planning to open the biggest new facility of its history—the $6 million World of Primates center that would finally release the primates from their embarrassing captivity in the outmoded old Rare Mammal House. The World of Primates should pack them in next summer. But this summer, the only thing to boost attendance was one little animal, the koala.

The more Biddle thought about it, the more it appeared that in fact the entire recovery of the last five years since Bill Donaldson took over seemed to be riding on the back of the koala, too. When Donaldson came on, the zoo was by many accounts ready for the garbage heap. Things were so bad, there was even talk of abandoning the site, which had been home to the zoo for over a century, and moving to the suburbs. The zoo's natural audience was being lured away. Television was showing animals stampeding, hunting, feasting, and doing other thrilling things that made the static zoo displays seem dull by comparison. Zoogoers were straying to sports stadiums, theme parks, and, most galling of all, those shopping malls. And the demographics were slipping. The baby boomers had grown past zoo age and hadn't yet produced another generation of zoo-goers.

Without municipal funding, the previous administration had tried to make up the attendance money by borrowing from the general maintenance funds, but that was like patching a hole in a sinking boat with a plank ripped from the hull. Buildings that were already starting to show their age five years ago, now looked as if they might fall down any minute, further scaring off potential visitors. But the zoo had shockingly few financial controls. In a development that seemed emblematic of that whole sloppy period, $120,000 worth of membership renewals disappeared one fall, only to bubble up out of an open manhole in an outlying section of Philadelphia the following spring. Apparently, the renewals officer had left the envelopes in his car overnight. The car had been stolen, and all the papers dumped into a stream by the thieves. The stream froze. The envelopes didn't reappear until the spring thaw.

When I went in to see Biddle about the koala, he was in a rage about parking. Parking was always a tender subject at the zoo because there was so little of it. On crowded days, zoo-goers had to park a quarter mile away and ride a bus back to the main gate. Now he was irked because he had paid $4,000 for directional signs to ease the flow of traffic into the parking lot by the front gate, only to find that the signs had been installed *pointing the wrong way.* "This does make me a little mad," he admitted. "I've said it all along, the key to the zoo is the first thing that happens. It's parking. Parking, parking, parking."

If the zoo couldn't handle the regular flow of visitors, he worried, how was it going to deal with the hordes coming in to see the koala? Biddle was counting on hordes, desperate for them. The zoo had budgeted 225,000 visitors, 75,000 more

than the usual August attendance. Biddle wanted 300,000—
150,000 extra. That would bring in $750,000 profit, a nice sum
in the zoo business. Biddle had earmarked the money for
planned improvements to the Bird House, a new picnic grove,
expanded winter quarters for delicate animals that couldn't stay
outside, a new upscale restaurant, and improved lighting. This
is what he had meant by the theory of momentum: Profits get
reinvested for greater profits in the future. Biddle wasn't a
banker; he was a builder.

In Biddle's heart he knew it was chancy, which made him
want it all the more. "You gotta take risks," he said. "That's
how you grow." Cincinnati's koala might have pulled in 100,000,
but that had been in June. This was August, and Philadelphia
in August is not pleasant—it is hot and humid. Philadelphians
evacuate the city in August as surely as Parisians, and for those
who stay, there are many other things to do in the heat besides
stand in line to see a koala.

And if he didn't get the crowds?

"If we're off, we've got to scale everything back—the picnic
grounds, the Bird House renovation, the winter quarters, the
north gate restaurant. Everything. Those are projects I'd really
like to have. If we can pull this off, we can get those things,
and we'll really have a better zoo. I really want to dig for this
one. I feel like going outside the gates and pulling the people
in myself. I'll use a bullhorn to get 'em. I'll go out there and
bellow, 'Come on in and see the koala! Come on in and see
the koala! Come on in and see the koala!' "

Biddle might have had another fear in the back of his mind,
but he could be excused for repressing it. When a venture banks
heavily on a risky proposition, a serious shortfall in revenues
can threaten more than some building plans. And if the zoo
itself, with its 111-year history, were to be shaken or—perish
the thought—go under while he was business manager . . .
well, that would be a disaster, nothing less.

—7—

The Philadelphia Zoo gets some quarrel over the title of America's first zoo from its colleagues at New York's Central Park Zoo, which had opened its gates thirteen years earlier, in 1861. But the Central Park's collection, never very impressive, was still so sparse as to constitute no more than a menagerie, and menageries had been around for some time. Bill Donaldson cattily dismisses Central Park's claim: "One swan on a pond and one bear on a chain do not a zoo make." The Philadelphia Zoo had actually been founded back in 1859, when a highborn Philadelphia physician named William Camac gathered some of his scientifically inclined friends to his house and drew up a charter of the Zoological Society of Philadelphia. Its intention was "the purchase and collection of living wild and other animals, for . . . the instruction and recreation of the people." This charter was approved by the state legislature in Harrisburg on March 21, 1859. But the outbreak of the Civil War kept the zoo from opening until 1874.

There had, in fact, been animals on display in Philadelphia since colonial times. "With the exception of an occasional hanging," wrote one chronicler, "our robust colonial ancestors had very little in the way of entertainment until the arrival of the menagerie." Hunters occasionally returned from the woods

79

with bears on chains to exhibit in city taverns; sailors brought back other exotics from across the sea. The city saw its first lion in August, 1727, when a ship captain put one on display at Abraham Bickley's store and charged viewers a shilling apiece. The beast then toured the countryside for eight years and became so well known to the local population that, years later, a newspaper could refer to it simply as "the Lyon." A camel showed up in 1740, then a moose, and later "a beautiful creature, but surprising fierce, called a leopard."

In the 1780s a precursor to the Philadelphia Zoo sprang up in back of Peale's Museum, the famous "repository of natural curiosities."

Charles Wilson Peale was probably the finest American portraitist of his day, but his work hadn't gone down well in Quaker Philadelphia, which dismissed his art as idle vanity. He had a small gallery in which to exhibit his paintings, however, and one day he displayed there some "mammoth bones" that a local naturalist had hired him to draw. The curiosity attracted quite a crowd. In 1786 he advertised for some more wonders. Benjamin Franklin sent him a dead French angora cat and George Washington some pheasant corpses. Peale sought to display the whole panoply of the natural world, properly classified by the Linnaean system. His most famous acquisition was the remains of two huge mammoths he pulled out of a bog in upstate New York. The bones helped disprove the argument put forth by some European naturalists that American animals were smaller than those of Europe. He inaugurated that display by inviting twelve guests to dine with him inside one of the skeletons; they toasted the great beast and sang a rousing chorus of "Yankee Doodle." To provide the rest of his mounted wildlife exhibits, Peale kept on hand a stock of live animals, eagles, owls, wildcats, even a baboon, which could be heard to shriek whenever Peale's mischievous assistant tossed him a quid of tobacco to chew on instead of the expected apple.

Peale regarded these animals only as the raw materials for his mountings, but noticed that others saw them as something more. When he moved into more spacious quarters back of the Philosophical Hall at Fifth and Chestnut streets, he added monkeys, panthers, a llama, an elk, an antelope, and several bears that were so ferocious one of them ripped a monkey's arm off at the shoulder in one swipe of its paw. Another broke out of its cage one evening and burst into the cellar of the Philosophical Hall, terrifying Peale's family, who lived upstairs. Peale calmly locked it in for the night and, in the light of morning, went down and shot it. He killed the bear's mate while he was at it, and mounted the two of them in an exhibit upstairs. He displayed a bald eagle with a sign, FEED ME WELL AND I'LL LIVE A HUNDRED YEARS. This proved to be an exaggeration. Stuffed, the eagle ultimately served as the model for the frieze over the main entrance at the U.S. House of Representatives in Washington.

Eventually, the museum wasted away—squeezed on the scientific side by scholarly organizations like the Academy of Natural Sciences, which offered similar mountings for free, and on the popular side by the likes of P. T. Barnum, who provided the public with ever more lurid spectacles without Peale's regard for scientific veracity.

It was decidedly on the scientific side of the equation that William Camac weighed in with his zoo a few years later. Camac (the emphasis is on the second syllable, to rhyme with attack) was not a great zoo man, on the order of William Temple Hornaday, who founded the New York Zoological Society (the Bronx Zoo) and ran it for thirty years on the force of his character. But Philadelphia didn't require that. The city was science mad, and its leading citizens sacrificed most of their leisure, and not a little of their money, to the craze. The cause could be traced back, like so many things Philadelphian, to the city's intellectual founding father, Benjamin Franklin. He had begun the American Philosophical Society in 1743 in imitation of London's Royal Society, for the promotion of "useful and practical" investigations into the physical sciences. This grand old institution spawned dozens of smaller ones; Philadelphians established scientific organizations and learned societies the way westerners started saloons. By Camac's time, the city supported the Horticultural Society, the Academy of Natural Sciences, the Franklin Institute, the Pennsylvania Historical Society, the Geographical Society, and the Genealogical Society, to name only the more prominent institutions.

While these organizations may sound today like generic

items of any civilized society, in the middle of the nineteenth century they were unsurpassed in America. And all of them bore an unmistakably Philadelphian stamp. When the members of the Academy of Natural Sciences threw a banquet in 1854, it consisted, according to Nathaniel Burt's *Perennial Philadelphians,* of "two fish dishes, four boiled meats, ten side dishes, five roasts, pheasant, prairie grouse, partridge, terrapin, fried oysters, six pastries, ten desserts, Madeira, champagne, pale sherry, claret, brown sherry, Shwartsberger, Steinberger, Liebfraumilch, brandy, whisky—and punch." Even though the academy's collection featured, as was often said, a "moose shot by a Biddle and stuffed by a Cadwalader," John James Audubon, for one, didn't feel that he had been fully accepted as a member of the American scientific establishment until he had been elected for membership in the Academy of Natural Sciences. This he finally succeeded in doing, after twice being blackballed, by publishing *Birds of America.*

William Camac was a member of the Academy of Natural Sciences, the Franklin Institute, and the Pennsylvania Historical Society. He never did practice medicine, although he was customarily addressed as Doctor. He was reportedly badgered into attending medical school by his fiancée, who insisted that she wouldn't marry him unless he had a profession. Wealthy from his family's land holdings in Ireland, he continued his scientific pursuits after obtaining his M.D. degree. He owned the first aquarium in Philadelphia, maintained a splendid conservatory filled with exotic plants at his country house, and he delighted in traveling abroad to examine the progress of science in other countries.

One of the scientific developments that interested him most was the London Zoological Society, which had been founded in 1826 as the world's first zoo. Here the animals were gathered for scientific observation rather than aesthetic decoration, as at a menagerie. Menageries, generally established to show the far-ranging dominion of a ruling potentate, had been around since Egypt's Queen Hatshepsut displayed monkeys, leopards, and a giraffe in the fifteenth century B.C. Although considerably expanded and improved since its founding, the London Zoo began life as a collection of simple, almost shed-like buildings in a garden setting. Later, in 1877, it inaugurated the word "zoo" into the language when it surfaced in a popular song with a strangely modern ring that included the line "Walking in the zoo is an OK thing to do." The Zoological Society was not, however, intended for the public. In their zoo prospectus, the society's founders, Sir Stamford Raffles and Sir Humphrey Davy, stressed that the animals were to be "objects of scientific research, not vulgar admiration." The sentiments must have pleased Dr. Camac. Initially, the London Zoo was open only to members and their guests, although the society relaxed on that point in 1847 when it offered tickets to non-members on weekdays for a shilling apiece. Three hundred thousand people turned out in 1850 to see London's first hippopotamus. (Only Thomas Macaulay seems to have come away disappointed. "I have seen the Hippo both asleep and awake," he declared, "and I can assure you that, asleep or awake, he is the ugliest of the works of God.")

Undoubtedly, Camac also paid a call on the Jardin des Plantes in Paris. Nominally a botanical garden, it came to have animals only by a curious set of circumstances. Like many rulers, the French kings had always had a great fondness for animals and had built aviaries, lion houses, and bear pits. Louis XI brought the canary to Europe. But Henri III nearly put an end to the royal collection in 1583 after a nightmare in which he saw wild animals tearing themselves to pieces. When he awoke, he had the entire collection shot dead. However, his descendants missed the animals, and gradually restocked the royal menagerie. It flourished under Louis XIV. He personally designed the Ménagerie du Parc at Versailles, in which, for the first time, all the animals were concentrated in a single fanlike exhibit rather than scattered about the landscape in the usual fashion. Louis also decorated the exhibits with flowers, bushes, and trees. Visiting monarchs rarely passed without a tour.

During the French Revolution, however, the populace was appalled to see royalty lavish attention on exotic pets while the common man starved, and outraged to hear tales of the keepers who squandered their daily six-bottle allotment of wine on the elephants and the camels for the fun of watching the beasts get tipsy. In response, the revolutionaries claimed the animal collection for the people.

It took the mob a while to get around to the Ménagerie du Parc, but finally in October 1789 it arrived at the gates to liberate the animals. The director of the menagerie, however, held his ground. He pointed out that some of the animals were extremely dangerous. Possibly they should be left in their cages for safekeeping? The crowd was slowed by this. The leaders deliberated. In the end, they decided to liberate a few token animals of the more peaceable variety. Consequently, as late as 1840, hunters were bagging Asian birds and some unusual goat species in the surrounding forest. The more fearsome animals in the collection were left where they were.

After the crowd departed, the director arranged to transport the remaining animals to the Jardin des Plantes for safekeeping. Formed as an herb garden where medical students could concoct their potions, the Jardin des Plantes had gradually evolved, by the time of the revolution, into a museum of natural history. Its director, fearing for his plantings, showed some reluctance to let animals onto the grounds, but he finally relented. And there, in the Jardin des Plantes, the animals have stayed ever since.

Inspired by these European zoos, Camac returned to Philadelphia with the idea of building one in his native city. There was no shortage of scientific interest in the project. Among the many Philadelphia dignitaries who turned out for the venture in 1859 were John Lawrence Le Conte, M.D., a preeminent coleopterist, or student of beetles and weevils, and John Cassin, who had written extensively on ornithology, mammology, conchology, and natural history. But even the hardiest industrialists and dullest-seeming lawyers proved to have a passionate interest in natural history. James C. Hand, the zoo's vice-president, was an iron merchant and early supporter of the Pennsylvania Railroad who also served as a member of the academy. William Parker Foulke wanted to mine the coal regions of Pennsylvania, but his academic interest in the subject was nearly as keen: He donated to the academy the first skeleton of a dinosaur to be found in America, a Hadrosaurus named Foulkei in his honor. The lawyer George Biddle had translated scientific works from the Greek. Dr. William Alexander Hammond was to be surgeon general of the United States and an early researcher into the physiology of sleep. And so it went.

Despite this impressive array of scientific dedication, the group had trouble coming up with the money. Then the Civil War broke out and ended their efforts for the time being.

Camac himself wrote an unsigned newspaper column after the war to try to revive interest in the subject of the zoo. "There is in our city, unfortunately, a great dearth of healthy forms of amusement offered to the public," he wrote in 1867, "and in looking over the statistics of foreign cities, we find that gardens of this kind [meaning zoos] have greater attention paid to them than any other places of recreation." In the spring of 1872, Camac put out a call to the twenty-seven surviving members of the original thirty-six founders to start up activities once more. Only seven showed up, but they were devoted to the cause. Putting up substantial sums of money themselves, they issued shares of stock for the rest at $100 a share and 6 percent interest, plus 2½ percent to be paid in the form of tickets to the zoo. Despite the Panic of 1873, which occurred in the middle of their fund drive, the zoo netted over $100,000. To cover operating expenses, the board decided to charge admission at twenty-five cents per adult, ten cents per child, a fee that would stay in effect for the next fifty years. (In 1985 the price was up to $4 for adults and $3 for children.) A Quaker—the only one, so far as it is known, to participate in the zoo's founding—named Alfred Cope offered a loan of $25,000 on the condition that "no vinous, malt or spiritous liquors be sold, and no circus or theatrical performances be allowed in the Garden." The board agreed, but the issue would nag at them for a century as liquor sales loomed as the key to profitability. They did succeed, however, in fighting off a proposal from certain religious groups that the zoo be closed on Sundays.

The Fairmount Park commissioners offered the zoo the strip of land on the western side of the Schuylkill River just below the Girard Avenue Bridge, which was then under construction and was destined to be for a time the widest (though far from the longest) bridge in the world. With the site came the historic house of William Penn's grandson, John Penn. William Penn had once owned all of Pennsylvania by a grant

from King Charles II, and he had designed Philadelphia's grid
street plan, but his holdings were seized during the American
Revolution. John Penn had sailed from England to try to re-
cover his grandfather's lost estates. Instead of reacquiring
Pennsylvania, however, he had to buy this little plot of fifteen
acres for £600 $terling. He had designed the house himself
and called it Solitude. Done in a Georgian style with a lime-
stone façade, the twenty-six by twenty-six-foot mansion is as
upright and sprightly-looking as a top hat. The *Architectural
Record* was once moved to call it an "exquisite box."

The first zoo managers put up a white picket fence to en-
close the property and hired the engineer Herman Schwartzmann, who later laid out the grounds of the Centennial
Exposition, to sketch a plan for the zoo. Then they dispatched
him to Europe to figure out what kinds of buildings to
erect on it.

They put out a call for animals, too. Dr. Camac enlisted
the aid of Frank J. Thompson, a dashing naturalist, explorer,
and adventurer then based in Australia, to be the superintendent of the garden, and to bring some animals with him. In
his letter to Thompson explaining why he could offer him no
more than a $2,000 salary, Camac declared, "You are aware
that it is a difficult matter in a country where these things [i.e.,
zoos] are strange and new, to make the people realize sufficiently the importance of them to come forward with the Al-
mighty Dollar." Keen on birds, Camac asked especially for
"plenty of cockatoos, parrots, flamingoes." Thompson had
camped by the falls of the Zambezi River before Livingston or
Stanley set foot in Africa, he had dug diamonds at Kimberly,
South Africa, and most important, he had captured wild animals alive. He brought back a variety of Australian mammals,
too, including a rare Tasmanian devil, although he lost several
more animals to injury in stormy seas. As a superintendent,
though, the jungle explorer left something to be desired—po-
liteness, most likely—and was fired after two years.

Other dignitaries responded to the call for animals. General William Tecumseh Sherman's wife sent Atlanta, the cow
that had accompanied her husband's march through Georgia
during the Civil War. Mrs. Sherman sweetly explained that she
"hoped the cow might pass the remainder of her life in peace."

A retired general from the U.S. Army sent grizzly bears, black bears, cinnamon bears, and brown bears to inhabit the zoo's bear pits. Brigham Young, the Mormon leader, added a couple of black bears, and the zoo gratefully made him an honorary member. The animals were transported free on the nation's fledgling railroad system, which meant, the directors soon realized, that when the beasts didn't make it back alive, the zoo couldn't very easily sue for damages. The railroads hadn't yet fully penetrated the western wilderness either. So, in general, the zoo preferred to transport its animals by ship. It reduced the expense of importing them by pushing a bill through Congress that allowed the duty-free importation of live animals for zoo exhibition. Consequently, the zoo's initial collection was more heavily weighted toward the animals of Australia and the Far East than those of the American West.

Frank Furness and George Hewitt were the most prominent architects whom the engineer Schwartzmann hired to put up the buildings he needed. Themselves old-line Philadelphians, they had built many of the mansions on Society Hill. The two were best known outside Philadelphia for their work on the Pennsylvania Academy of the Fine Arts, an extraordinary work that became a kind of tribute to industrialism, with columns in the shape of pistons. By a quirk, the zoo hired them at a point in their architectural development when they were engaging in the highest flights of fancy. Frank Furness did the rustic gatehouse, which, true to the raging gingerbread style, looks like the witch's house in Hansel and Gretel. Furness and Hewitt together designed the equally gay Antelope House and a colossal elephant house with square towers and a timbered upper story. Inside, the stalls were elevated, allowing the beasts to look down on the visitors as they passed through the central walkway.

The zoo opened on schedule on July 1, 1874, and as the brass band played and banners flapped on the masts out front, visitors streamed through Furness and Hewitt's gatehouse in a quantity sure to comfort the zoo's financial directors. Six hundred and sixteen animals were on exhibit, including antelope, elk, prairie dogs, zebras, kangaroos, an elephant, a rhinoceros, and a tiger. The zoo's initial collection of six giraffes would not be equaled for fifty years. Six snakes were billeted in the library at Solitude; the keeper lived upstairs. Some of the more memorable names for the animals reflected the continuing struggle over Darwinism. A pair of chimps were called Adam and Eve, and a baboon was named Henry Ward Beecher after the minister and evolutionist who had just been accused of committing adultery with a member of his congregation. A reporter from *The Age* seized on the social Darwinist implications when he toured the Monkey House. "There are some fifty of these little fellows [monkeys], jumping about their cages, running up and down the swing ladders," he wrote, "as if in constant dissatisfaction with their positions in life, taking almost as little time for recreation as their great descendants in the struggles for position on the uncertain ladders of wealth and fame."

The elephant, Jennie, caused the most excitement on that opening day. Apparently, the chain securing her to a stout oak was not visible in the tall grass. A "middle-aged maiden," according to one newspaper account, took such a fright when the beast strolled toward her that she dropped her parasol and tried to climb a tree. Only then did she notice the chain. That was only embarrassing. The day's tragedy occurred when a kangaroo was spooked by the sound of a train passing on the Pennsylvania Railroad tracks behind the zoo. It ran into a fence and broke its leg. Other mishaps were less innocent, however. Two weeks after the opening, the zoo announced that a South American sloth had been poked to death by umbrellas and canes, and that a baboon—it is not recorded if this was Mr. Beecher—had been killed after being fed sulfur match heads.

But *Harper's Magazine* summed up the general feeling in an adulatory article shortly after the opening that declared the zoo was "lacking hardly anything of grand importance to the mass of patrons, unless we might mention the hippotamus."

A quarter of a million visitors turned out that year, and nearly 450,000 the next, well exceeding the numbers for the London Zoo. In 1876, the year of the great Centennial Exposition, the zoo attracted an astounding 675,000, a figure that would not again be reached until the opening of the Children's Zoo in 1955. Ulysses S. Grant came to the exposition and while he was there, looked in on a pair of cassowaries, large, flightless birds from Australia, which he had sent the zoo some months before.

The annual attendance slid back to the two hundred thousands for the rest of the decade; and, further eroding revenues, members took to scalping their free tickets. The zoo continued to build its animal collection, however, most significantly with the addition of Pete, an Indian rhinoceros purchased from P.T. Barnum's circus. Barnum was willing to part with Pete after he developed a habit of rocking back and forth in his cage wagon, nearly yanking the horses out of their harnesses. He would become a zoo fixture until his death in 1901, whereupon he was delivered to the Museum of Natural Sciences across the river and stuffed. That was a common fate for zoo animals in those days, and it put the zoo in the old position of the menagerie in back of Peale's Museum so many years earlier. There continued to be a good-natured rivalry between the proponents of living animals and the enthusiasts of mounted exhibits. Williams Biddle Cadwalader, a professor of neurology at the University of Pennsylvania, came to run the zoo at the start of the century; his brother, Lambert, was a major patron of the Museum of Natural History. At the Philadelphia Club, where the brothers often lunched together to argue the respective merits of the two institutions, the two gentlemen became known as "The Quick and the Dead."

Dr. Camac took his leave from the presidency in 1878, and from the zoo altogether a year later. He came to a sorry end. Some years later, he took his family and a retinue of servants to sail by houseboat up the Nile. He left his financial affairs in the hands of a friend back home. Unfortunately, the friend absconded with the money, leaving Camac virtually penniless on his return. Camac left Philadelphia and moved to New York. Today, he is nearly unknown in the city. The situation is little better at the zoo. His portrait hangs with those of the rest of the directors along one corridor of the administration building, but otherwise there is no memorial. None of the buildings are named after him; no sculpture of him adorns the grounds. At the zoo he founded, he is a forgotten man.

8

Chuck Rogers says that Philadelphia falls into the "mean" temperate zone, which only sounds mild. The city has arctic winters and tropical summers, with brief periods of pleasantness in between. Its climate ranges the globe, that is, and some summers descend into hell. By late July, as the zoo waited for K'bluey, the fruit trees had long since dropped their lovely scented blossoms. The rhododenrons and azaleas had wilted, and, with the mercury up around ninety, the only vegetation making much of a go of it were the cacti.

All the animals looked a little groggy in the heat, and they moved to a largo beat. The female hippo, Subbie (for Submarie), outside the elephant house was spending nearly all her time under water. Her consort, Frankie, stayed indoors. The zoo used to keep them together until one time Frankie got so excited during one sexual escapade that he nearly bit Subbie's head off. To tell the truth, zoo staffers would not have been terribly sorry if he had succeeded. It takes a lot to love a hippo with its Zeppelin body, stubby legs, and bulging snout. Hippos have rotten tempers; they are so lazy that they try to intimidate their rivals by yawning; they take up a lot of space; and they breed so freely that their market value is nil—zoos can't give them away. "Big blobs," says Dave Wood. "There's not much personality there."

Down the walk, outside the lion house, the African lion, Webster, never very active, only flicked the tip of his tail to show that he was still alive as he stirred in his slumbers on his rocky domain under a locust. Off in Bear Country, the polar bears, with no snow to eat to cool themselves off, took refreshing plunges into their swimming pool. Unable to shed the thick layer of blubber that kept them warm in arctic temperatures, they panted like dogs the rest of the time. The flamingoes always used to fade in the bright summer sun until the former curator of birds, Gus Griswold, figured out that they would keep their color if he fed them carrot juice with their dinner. Now they stayed fluorescent pink all summer long. The rhinos, Billy and his pregnant mate, Xavira, were used to African standards of heat, and sensibly moved only to follow the shade of the maples and the sycamores across their enclosure. Outside the Rare Mammal House, the older gorillas picked idly at their fur while the youngsters cavorted about them, oblivious to the scalding temperatures.

The animals just in from the Southern Hemisphere had the worst of it, because they were still stuck in their thick midwinter coats. A Bactrian camel from Australia that the camel-ride concessionaires Tim and Tracy Hendrickson had acquired was in that predicament; and you could tell from the itchy way she walked that she couldn't wait to take her coat off. Tim and Tracy showed little pity except to enforce a four-hundred-pound limit on the riders.

The only animals that couldn't take the Philadelphia summers were the reindeer. It wasn't the heat but the humidity that did them in. So the zoo unloaded them.

Only Mopey-Dopey, the Galápagos tortoise, seemed to have much fun. All summer long he was sawing away on somebody—or something. He was pushing sixty-five, as near as could be determined, and he'd been at the zoo a good twenty years, but his age hadn't diminished his ardor, just his eyesight, and possibly his brainpower. Several boulders in his yard were the objects of his affection. So was the Reptile House's front doorstep. And I thought I once caught him casting a lascivious eye at the back fence. Whatever his fancy, he was up and at it all day long like a teenager, and you could hear him from the four corners of the zoo. *Grrrrrruuuugh, grrrruuuuuugggghhh, gggrrrrruuuuuugggghhhhh.* He sounded like a chain saw in low gear—a low, rumbling, satisfied sound.

While the rest of the pack in the tortoise yard bore all the intelligence of the inverted soup bowls they resembled, Mopey-Dopey obviously had a mind, even if there was only one thing on it. He didn't mount his love mates so much as he scaled them. First he gained a toehold; then, with a groan or two, he shoved himself up on their backsides; then he dropped on them; and then he felt around with his member for their tender parts. Kids were invariably drawn to the spectacle and gaped open-mouthed from the fence. They didn't always get the right idea, however, and I heard many a shriek—"Hey, Mommy, look! They're playing horsey!"

It must have puzzled Mopey-Dopey to find that he had indeed mounted a rock, but in truth his other lady friends weren't much more responsive. They generally just lay there thinking about dinner and waiting for Mopey-Dopey to have done with them. Sometimes, the ladies in question were barely half his size, so he was able to climb up on them, then loop his head back upside down to stare at them beak to beak, gazing deep into their eyes, as if to ask, "Fair maiden, how like you this?"

Mopey-Dopey was so preoccupied that he missed the big breakout. I came by the tortoise pen one afternoon to see Denise Robinson, one of the new wave of keepers with a master's degree in biology, and I heard Mopey-Dopey going at it, so I assumed that all was well. I was talking to Denise inside the Reptile House when another keeper came running up, screaming for Denise's help. The two of them raced outside. And I dashed after them.

Prickles was loose! A four-hundred-pound, middle-aged male tortoise, Prickles had little of Mopey-Dopey's intellectual prowess. He'd slipped out, but had entered into the land of the spectacled caimans—those nasty six-foot-long crocodiles from Paraguay that Dan Maloney kept as pets—next door. It was like escaping Devil's Island by jumping into a sea full of sharks. Apparently, Prickles had been rumbling along the fence and happened to bump against the unlocked gate into the caiman yard. It is hard to believe that this feat was premeditated; Prickles was almost certainly just being clumsy. In any case, the gate swung open, and Prickles lumbered ahead. If he felt any sense of liberation after a decade of confinement, he didn't show it. Fifteen feet from the open gate, Mopey-Dopey kept his mind on business, and continued to thrust away without so much as looking up.

The caimans, like most of the crocodilians at the zoo, normally spend their time doing log imitations in the pool, but the keepers feared that their behavior might well change if Prickles accidentally stepped on one of them. It was an open question whether he would squash the caiman or merely irk it sufficiently to snap its jaws on his leg. Rather than wait for the answer, the keepers had appeared with a small arsenal of scrub brushes and shovels to try to persuade Prickles to turn around and go back. The keepers started out gingerly, giving Prickles love pats with the scrub brushes on his rump.

When Denise arrived, the keepers started whaling away at his backside with the shovels. Each blow made a dull, whumping sound as though it were striking heavy cardboard, and the sound combined intriguingly with the grunts of ecstasy emanating from Mopey-Dopey next door. The blows got Prickles moving, but in the wrong direction. So one of the keepers dashed back indoors and came back with a sheet of plywood. If the blows to his backside were the accelerator, the plywood was the steering wheel. Steering Prickles, however, was like trying to steer the *Queen Mary*, the tortoise had such a wide turning radius. In order for the tortoise to be moved back to the gate, Prickles had to swing around by the caimans' pool; he accidentally splashed a foot down in caiman waters as he passed. Remarkably, the caimans didn't notice—or at least *pretended* not to notice—but continued to float in the water with just the tips of their snouts and their beady eyes in the air. When Prickles squeezed back through the gate and into the tortoise yard, the keepers broke out into wide, relieved smiles. Mopey-Dopey was still grunting away, still hard at it.

All the animals may appear to be little more mindful of their environment than Mopey-Dopey, and yet, on a keen subliminal level, some of them at least know what's going on. They know the rhythms of the zoo. They know how far the shadows fall before it is time to go in. The springbok and blesbok, the slender antelope that bound across the lawn at African Plains, for instance, gather about the gate with increasing impatience as the magic hour of four o'clock draws nigh. And across the walkway, the giraffes, always fidgety, start to run their noses along the doorway, eager for their dinner. It had taken a good three weeks to get the animals adjusted to the change from eastern standard time to daylight saving back in April; and no doubt it would take another three weeks for them to turn their internal clocks back again in October.

The animals are attuned to subtler senses of order as well. Never had the zoo heard such a fuss as when the monorail, which for the last ten years had made regular clockwise trips around the zoo, accidentally started to go *counter*clockwise. The camels panicked and nearly threw their riders. The elephants stampeded about the yard, their trunks upraised, and trumpeted furiously as if there had been an earthquake. All around the zoo there was such a frightful clamor, it took the keepers nearly an hour to quiet all the animals down.

Most of the animals have a dim sense of the staff, too. Elephants and gorillas are by far the shrewdest. They are able to pick their keepers out of a crowd. The gorillas can even do it when the keepers are in civilian clothes. And they remember their faces for years afterward. After Snickers had her baby, she brought it to the glass to show Dr. Snyder when she spotted him walking by; Snyder had raised Snickers ten years before.

All the mammals seem to know which keepers they can take advantage of and which they can't. And all but the very stupidest—like tortoises and hippos—know to be wary of the veterinarians, Keith Hinshaw and Mike Barrie, and to watch out when the keepers' blue van is headed their way. The snow leopard had to be knocked down with the blowgun every Tuesday during the summer so that the vets could check her sore paw. No fool, she started to snarl and stalk angrily about her cage as soon as Keith and Mike stepped into the building. Some animals carry sullen prejudices of their own into the zoo. Wolves would sometimes take an intense dislike to a visitor for no apparent reason; they'd bore into that person with their eyes and start howling to bring the rain. Surprisingly often, the security guard end up escorting those people to the gate before their visit is complete. Troublemakers.

Aware as the animals are of their environment, it is hard to know how confined they feel by their cages and enclosures. It is common for outsiders to decry zoos as "prisons" for animals, but few zoologists believe that the animals themselves see it that way. Deeply as Americans cherish the notion of liberty, animals have no use for it. If anything, prison might well strike them as positively desirable if it meant that food was plentiful and they were safe from predators. Territorial by nature, animals in the wild tend to stay within remarkably tight boundaries where they can be assured of a food source. Birdsong, for example, is not the lusty cry of the free at heart, but rather the defensive shriek of the proud homeowner, saying over and over, "This space is *mine*"; it is also an advertisement for mates.

At the zoo the animals spend hours laying down the scent markers that act as the olfactory equivalent of PRIVATE PROPERTY signs: Siberian tigers do it with an aerosol spray of urine; slow lorises, skinny, ratlike primates with bulging black eyes, have a gooey pad inside their forearms which they wipe on bars and branches; binturongs lay down a scent with a swipe of their anal glands. By these means, the animals make the space into their territories. Indeed, one of the animals' greatest frustrations in life is the fact that their keepers come in regularly to clean—forcing them to scurry around and mark their space all over again. Such cleaning is almost cruel; it is as if someone had rearranged the furniture in a blind person's house.

Even when the animals have the opportunity to escape, they rarely take advantage of it. The tahrs, husky mountain goats used to scampering about the Himalayas, could easily hop over the five-foot fence that ostensibly keeps them penned in their rocky yard. Fantastic leapers, they can bound ten feet into the air. Sure enough, they have hopped out a few times— but they always hop right back. And the gawky secretary birds roost in branches that overhang the fence to their enclosure, but they know to stay inside. When a herd of roe deer was accidentally let free from the Bern Zoo in Switzerland a few years ago, the deer immediately took off into the neighboring forest. But within a few days they were all back at the zoo gates once more, eager to be let in.

Of course, great escapes do occur. One of the memorable ones took place in 1940, when a passel of the Philadelphia's rhesus monkeys broke loose from Monkey Island and terrified the local citizenry. The island had been opened only a few days before and had been dedicated by the jungle explorer Frank Buck. There had been a terrific May Day celebration, and the keepers drained the water from the moat around the island to clean out some debris from the celebration. Unfortunately, nobody thought to fill it back up again. Fifteen monkeys raced across the empty moat, bounded over the wall and out to the street. There they scrambled over parked cars, climbed up drainpipes, and danced around the rooftops. Some swung by their tails from the eaves; others peeked in windows. The residents went wild with fright. One monkey was incinerated when it fell down a chimney into a burning fireplace. Another burst into a taproom, to the astonishment of the patrons. And a third caused a young lady named Marie Charleston to faint dead away when it dropped from a telephone pole onto her shoulder. The zookeepers finally managed to round up the escapees by bagging them with butterfly nets, much to everyone's relief.

Aside from the fact that these are familiar territories, the animals have other reasons to enjoy their zoo homes. Despite the romance of the wild, the zoo has several undeniable advantages. To illustrate this point, Dr. Robert Snyder gives the case of the humble woodchuck. The director of the zoo's Penrose Laboratory, a small research facility located on the grounds, Snyder has maintained a side interest in woodchucks that is awesome to contemplate. It started when as a young biologist he was called in by the army to study an arms depot in central Pennsylvania that was being overrun by several thousand woodchucks. Since then he has studied five thousand living woodchucks and one thousand more dead ones. For the last thirty years he has used the woodchuck to investigate the role of the hepatitus B virus in the formation of liver cancer; the woodchuck is one of the few animals, aside from man, that are susceptible to the disease. Bill Donaldson kidded Snyder that he was even coming to look like a woodchuck, with his scrunched-up face and buckteeth, but that may be taking matters a little far. Although gardeners who have seen their vegetables disappear down woodchuck gullets might disagree, Snyder believes that these animals don't really have it so good out in the wild. From birth, they are vulnerable to a variety of predators. Mites and ticks wriggle into their pores and consign them to a lifetime of futile scratching; and woodchucks are likely to play host to tapeworms that feast on them from the inside as well. If they survive such hazards, and live six months till sexual maturity, then the woodchucks face an even sterner test—fighting it out with each other for a mate. The dating game is literally murder for woodchucks, for the losers in the gladiatorial combat for a spouse often bleed to death from their wounds, including the final indignity of having their tails plucked out by their victorious rivals as they flee. For all these reasons, the average woodchuck is a measly specimen—horribly malnourished, vermin infested, and ready to die by age one. By contrast, the woodchucks that are raised in the zoo in small metal cages off exhibit in the back of the Penrose Annex live up to twelve years. Despite the accommodations, they are twelve pretty good years—no predators to contend with, no diseases or pestilence, plenty of food, the best medical care, and all the sex they want. In fact, they live so long that some of them ultimately fall victim to such human ailments as lung cancer from Philadelphia's air pollution. All they lack is freedom.

Nevertheless, Snyder recognizes that any old cage does not suffice. Animals are like humans. They like to set aside certain areas of their territory for different activities: one spot for resting, another for feeding, another for defecation, and another for bathing. They need the variety. As the noted zoo authority Heini Hediger points out in *Man and Animal in the Zoo*, it can be far more deleterious to curb an animal's behavior than to limit its space. Beavers, for example, would die without a bathing pool, for they generally defecate only in water. Similarly, elephants have to bathe to keep their skin from turning dry and scaly. And other apparently playful behavior is as basic to animals' nature as their size or markings. Monkeys need to climb; armadillos need to dig. Beyond this, Snyder believes that all animals understandably need to have a space into which they can retreat, out of sight. The environment has to provide challenges to keep the animal from getting bored. The higher the order of animal, the greater the degree of complexity needed. Still, the complication can involve anything from giving the chimps cardboard boxes to play with, to sprinkling sunflower seeds in the straw of the gorilla cage so the gorillas have to work to pick out their food.

Otherwise, says Snyder, the animals may not only grow listless and depressed but they might suffer from adrenal atrophy, meaning that their bodies might no longer be able to tolerate the burst of adrenaline that comes from sudden excitement. Snyder started thinking about the problem with particular urgency back in the 1950s when the cages in the Carnivora House were so utterly plain that the animals had nothing to do but pace back and forth all day. One morning, Snyder got a frantic call from the head keeper of the cat house, who shouted that seven of his cats, including three tigers, had just dropped dead—keeled right over for no apparent reason. Snyder raced over to investigate and found some workmen had been painting over the skylights above the cages to cut down the heat and light during the summer season. Apparently, the sight of the men up on the roof had given the animals the shock of their lives. Their enfeebled hearts couldn't take it; all seven of them had suffered massive heart attacks in seconds.

Snyder, however, presents the rosy view. Others at the zoo are not so sanguine about the zoo's effect on the animals it harbors. They don't weave broad theories, but rather point to specific animals that are clearly not thriving—and have persistent medical complaints to show for it. Some of them are self-induced out of boredom. For as long as anyone can remember, one spectacled bear had an open sore on one knee that he repeatedly rubbed against a rock. The African lioness Gira had a sore at the base of her tail that she wouldn't let alone. Then there is a malady that is gently termed hypersexuality. The drill Wilbur, named after the zoo's second-in-command Wilbur Amand, enjoyed public masturbation. More embarrassing to the zoo, he was somehow able to tell which visitors were going to be most affronted by the spectacle and singled them out—mainly nuns and the Amish—to witness his performance.

Another primate, the orangutan Bong, was a sadder case. As a youth he had been reasonably active, and produced five young. But then his mate, Christine, died, and he went into prolonged mourning. He sat on a limb of a Gunite tree in his exhibit and rarely moved a muscle except to breathe and, occasionally, to eat. He looked like a vast kitchen mop—a great mound of reddish hair, with a puffy face that seemed to exaggerate all the sadness that was in him. When Bong's keepers heard that a graduate student was going to come in and do a study of Bong, they recommended that the student try a boulder instead. Sure enough, the student abandoned the task after a week, having been unable to fill up a single page of his notebook with notes on Bong's behavior.

But Bong, happily, was the exception. Most of the animals at the zoo were doing fine.

——9——

It was a sunny July 29, and K'bluey was due in the lion's den any minute. Inside the entrance to the Carnivora House, the zoo had set up an imitation eucalyptus tree for the koala in a glassed-in, air-conditioned exhibit. If the building were the U.S. Capitol, this area would be the rotunda, with the lions in the Senate and the tigers in the House, and a variety of other cats— snow leopards, black panthers—lining the hallways. The cats must have envied the attention. Posters of K'bluey were taped to the glass. Afraid that K'bluey might choose to fall asleep in the back of the exhibit behind the foliage, workmen had rigged up the whole contraption on a rotating platform, a kind of enormous lazy Susan, so that he could still be swung around into view. Inside, the exhibit was soundproof, so K'bluey wouldn't hear Webster's thunderous roar down the hall. A koala, which has no natural predators, might not have known what to make of that.

K'bluey was flying in to the Philadelphia airport first class, courtesy of American Airlines. He had his own seat, but couldn't appreciate it because he was confined to a Sky Kennel. I learned later that he had nevertheless been able to enjoy the view and that he spent most of the trip gazing out the window. His keeper from the San Diego Zoo, Elaine Chu, sat beside him to keep him company.

The zoo had considered sending a limo, but for economy's sake ended up dispatching a regular zoo van to collect K'bluey and Chu at the airport. Now word filtered in that the pickup had gone smoothly, and the van was on its way back to the zoo.

Several newspaper reporters and radio people, plus crews from all three local news programs, were milling around outside the Carnivora House. The antennae masts on the back of the news vans were raised to beam the first reports of K'bluey's arrival back to their stations. The reporters didn't seem too happy about the assignment, though. "This is a yawn," said a grizzled cameraman for Channel 10. "It's noncontroversial. A koala comes to the Philly Zoo. That's not your four-alarm fire. We're advertising the Philly Zoo. We're promoting its new product, this koala. If Gillette came out with a new shaver, they'd have to pay to put it on TV. All we're doing is doing it for free." Some of his cohorts mumbled approval. Then he was called away to start taping. I edged closer to hear the report. Despite the cameraman's glum assessment, John Blount, the personable reporter for Channel 10, played it cute. "A koala is a marsupial, not a bear," he told the camera. "For a marsupial, he's got a pretty good deal—a twenty-four-year-old traveling companion, and he's only one. He's doing pretty good."

The zoo is custom-made for the media, and vice versa. "Zoo pictures are an easy smile," said one AP photographer, Amy Sancetta, who came out frequently to photograph zoo births. The Philadelphia *Inquirer* ran regular reports from the zoo, although it preferred stories with a scientific angle; the tabloid Philadelphia *Daily News* leaned toward soft features with lots of photography. The *Daily News* had the "gorilla editor," Frank Dougherty. Local TV stations relied on the zoo for upbeat footage. Even on days when there were no news stories, the local TV stations often dispatched vans to the zoo for footage to run in the background during the weather segment.

But even if some cameramen disagreed, the koala was big news, if only because it was such a popular animal. Why was a koala so popular? The geologist and science writer Stephen Jay Gould offers the best explanation. Building on work pioneered by Konrad Lorenz, he argues that humans are drawn to the animals that most resemble human infants, and he uses Mickey Mouse as a prime example. He notes that Mickey has changed over the years as his makers groped to find an audience. Compared to Mickey's first incarnation as *Steamboat Willie* in 1928, his features have softened noticeably. His nose has shortened, his head ballooned, his legs shrunk, and his eyes widened. In other words, he has grown more babylike. Humans invariably open their hearts to babies. Gould cites Lorenz in identifying the key features to elicit this response: "a relatively large head, predominance of the brain capsule, large and low-lying eyes, bulging cheek region, short and thick extremities, a springy, elastic consistency, and clumsy movement." All of this describes a koala except for the part about the eyes. For the koala's eyes are rather small, although they are widely enough spaced to seem bigger. (The panda, however, is a perfect fit, which may account for its superior popularity.) But the point is clear. When we see a koala, we see a lovely little baby.

"Here they come!" somebody yelled, and all the newshounds started running. The blue zoo van was creeping up the walkway. Wilbur Amand, the man in charge of the collection, was in the front seat next to the driver, his head turned around to keep a close watch over his cargo. The van stopped, and Steve VanderMeer, the head of the Koala Kommittee, popped out of the back along with Elaine Chu, a brisk young woman in San Diego Zoo khakis. Light bulbs started to flash as two staffers pulled out a large hard-plastic cage. The two carried it like a fish tank, careful to keep it level. Cameras were firing all around them. Their repeating mechanisms sounded like gunshots. The bright klieg lights for the TV cameras were going, too, and throwing the stark shadows that you sometimes see behind celebrities going to some gala on television. As K'bluey was carried past, I peered in the wire-mesh windows. I could see him huddled inside like a cat, clinging to a eucalyptus stump. The lights shone in his almond eyes.

The workmen set the cage down inside the doors to the Carnivora House, and the photographers clamored for Elaine to take K'bluey out for pictures. She pulled on a falconer's heavy leather glove that reached clear to her shoulder. She got down on her hands and knees, and stretched her arm in farther and farther, groping obviously for an animal that would rather stay where it was. "Come on," said Elaine in that breathy singsong that always spells trouble for animals, "come on, boy." K'bluey must have been hanging onto his eucalyptus stump for dear life. The photographers grew impatient.

Then Chu snagged him. K'bluey emerged in the crook of her arm. There was another explosion of light, and he peered around the room blinking. He sure was cute. He had fuzzy Mickey Mouse-cap ears, sleepy eyes, a big black honker of a nose, and a huggable little baby body. Gould was right. In truth, as Chu's thick glove attested, cuddling was out of the question. The apparently soft fur was actually rough as a Brillo pad, and the claws were sharp. Nevertheless, from all around the room came one sound in unison: "Awwwwww."

"Hold him up higher!" somebody yelled.

"This is as high as I'm going to go," she replied sternly. K'bluey sat in the crook of Chu's arm as though he had been born to it. Everyone else was scrambling—for a better look, to take notes, to get the right angle for the photo. K'bluey was the calmest animal in the room.

Closely following the San Diego Zoo's guidelines, a craftsman named George Chestnut had wired some maple branches together to look like the eucalyptus tree that makes up the koala's native habitat, allowing a variety of perches for K'bluey's resting pleasure (not to be overlooked since koalas sleep all but four hours out of the day), and he attached a few buckets here and there to the trunks to hold his eucalyptus branches. Koalas dine exclusively on eucalyptus; they die without it. It is not too much to call them eucalyptus-eating machines. They live in eucalyptus trees; their hands and feet are oddly splayed to cling to eucalyptus limbs; and their appendixes are eight feet long in order to process the tough eucalyptus leaves (the human appendix, by contrast, is about one inch). The eucalyptus diet is so hard to get used to that koala mothers wean their young on chewed eucalyptus leaves diluted in mother's milk, a substance excreted from the anus. So that the young can reach this food, koala pouches open backward. The animals' noses are refined to detect the extremely subtle fragrance of deadly prussic acid, which is sometimes present in the leaves. (Our noses miss the acid smell; we can detect only the presence of the koala, which, because of the eucalyptus, smells faintly like a throat lozenge.) Presented with a dozen eucalyptus branches, the koala might well discard eleven of them because they don't smell quite right. About the only design element in a koala that is not geared to eucalyptus consumption is the two fused fingers on each hand. This double claw makes a handy comb.

Eucalyptus doesn't grow in Philadelphia, which is why Philadelphia had never had a koala before. In the United States it grows only in California, so the San Diego Zoo had arranged to fly out a fresh supply twice a week. In order that the airlines would treat it with the respect it deserved, the box was marked "Orchids."

The photo session over, Elaine put K'bluey down in his new cage. As we watched, the koala entered it tentatively, shuffling along unsteadily, and sniffed around the inside edges. Deciding the place was okay, he bounded onto the tree and scampered to the very top of one branch. He stayed up there for a moment swaying a little like a sailor on lookout at the top of the mast.

Elaine now discovered the lazy Susan.

"Hey," she said to Bill Maloney. "We can't have that. I don't want him on any rotisserie."

Chastened, Maloney said he'd nail that down as soon as he could.

Clumsy as K'bluey had seemed on the ground, he was amazingly graceful and surefooted among the branches. As if to express his delight to be out of his Sky Kennel, he jumped from one branch to another, competent as any monkey.

Then—kablooey. Nobody had told him about glass. Apparently thinking that the glass's vertical support was as graspable as the trees, he leaped out and tried to encircle it with his front paws. But the glass got in the way. We heard a smack, then a desperate scratching sound as his claws struggled to gain a foothold in the glass. Down, down, down he went, and finally hit the floor with a heart-stopping thud. Everyone gasped to see K'bluey prostrate on the floor.

Elaine stepped quickly inside, picked him up, and set him back on his tree again. He huddled there, trying to clear his head, then ventured back up the tree. It looked like he was all right.

"I about had a heart attack," exclaimed Elaine when she reemerged.

Bill Maloney taped the glass so K'bluey wouldn't make that mistake again. But in case he did, Maloney also brought a couple of mattresses and laid them on the floor to cushion any fall. Someone brought K'bluey a snack of some fresh eucalyptus to cheer him up, and he sniffed a branch with interest. "Mmmm, lunch," said veterinarian Keith Hinshaw, who was watching. K'bluey started to eat. Elaine drew the curtains around the exhibit to let K'bluey nosh in comfort and then take a nap. The show over, the crowds dispersed

Koalas first came to the United States for a movie. That was in 1952, when Paramount wanted to film an account of the Australian penal colonies. The koalas supplied the local color. When Belle Benchley, then the director of the San Diego Zoo, heard about the animals, she asked if she could keep them after the moviemakers were finished. San Diego has a climate similar to Australia's, and it had an ample supply of the necessary eucalyptus. At that point, koalas were under the strict control of the Australian government, which granted its approval. The clearance procedure was cumbersome, but the government's intervention probably saved the koalas from extinction. For in the early part of this century, koala pelts— valued for their warmth—were all the rage, and up to a million and a half koalas were shot out of eucalyptus trees every year and turned into coats, hats, and handbags. It took special legislation to stop the slaughter.

The first European to see a koala is reputed to have been a Frenchman, identified only as Ensign F. Barallier, who in 1802 traded several spears and one tomahawk to the aborigines for "parts of a monkey which they called Colo." Unfortunately, Ensign Barallier was only able to obtain a single pair of koala feet, which had been shipped inside a bottle of cognac to the governor of New South Wales. Not until the end of the century did the first koala reach Europe intact. The London Zoo obtained this first specimen from a dealer in 1880 along with a fair amount of dried eucalyptus leaves. Remarkably, the koala survived this diet, and only died by accident when it hit its head on a washbasin. More pathetically, a sailor brought a couple of koalas to the London Zoo in a sack in 1927. Yet he had only included a small supply of eucalyptus. Newspapers printed desperate pleas for the precious leaves, but none were brought forth, and the animals starved to death in a month.

It is in some ways remarkable that humans nowadays are so devoted to these animals even with their similarity to human infants. Malcolm Smith, an Australian zoologist at the University of Queensland, spent 739 hours watching 113 koalas at Lone Pine Sanctuary in Brisbane, and determined conclusively that koalas are the laziest creatures on earth. Koalas do only three things with any regularity—eat, sleep, and scratch. Only rarely do they even bother to shift their weight. They build no nests, engage in precious little interaction with their fellow koalas, and do not have to hunt for food. Koalas sleep twenty hours out of every twenty-four, and they rest most of the remaining four. Koalas are so lethargic, in fact, that it was once thought that eucalyptus acted as a narcotic on them, but that suggestion has been discounted. The reason that koalas are such sleepyheads now appears to be that they have nothing better to do. Their food is taken care of, and man is their only predator. They probably couldn't do very much about us anyway. So they sleep.

Although Smith's research is couched in the dry, analytic terms of the professional scientist, he is plainly amazed that the koalas show so little curiosity about life around them. Strange as it must have been for them to find Smith there day after day, watching them with notebook in hand, they couldn't have cared less. One adventurous soul went so far as to sniff Smith's briefcase and take a closer look at his shoes, but that was it. All the others carried on as usual—which is to say, they continued to doze. Smith tried to arouse the koalas' curiosity by leaving unspecified "foreign objects" in their path. But they paid no attention. He would dearly have loved to quantify that attention, perhaps draw up a nifty chart like the one he used to detail the koalas' eating pattern, displaying the number of bites, the orientation of the bites, the amount of sniffing, the amount of steadying, the style of grasping, etc., but he writes, "The data was too small."

Despite all the affection we lavish on them, koalas are on a take-it-or-leave-it basis with humans. However, an Australian woman named Mrs. Oswin Roberts succeeded once in getting a koala to take to her in most literal terms. According to a fascinating little book called *The Cry of the Koala* by Ambrose Pratt, Mrs. Roberts had been watching the koala colony in the manna-gum trees behind her house on Phillip Island in Victoria when she noticed that one of the little charmers had no mother. She took the little fellow in, christened him Edward Lewis Oswin Roberts, Esquire—the "Lewis" was for the local wildlife official who allowed her to keep him—and fed him milk from a spoon. Edward took to the regime so gratefully that he treated Mrs. Roberts like his mother; that is, he climbed on top of her head and would not come down except to sleep. Even then, the first thing he would do upon awaking was to toddle over to her bed and snuggle under her chin. When she got up, he would lazily watch her dress, then climb up her as though she were a tree, and take up his position on her head, a koala chapeau. He ate his meals with the family, sitting in a high chair. As he aged, he grew bolder and used the living room as his own private eucalyptus arbor; he jumped merrily from chair to chair, ceasing only when Mrs. Roberts came into the room. Then he would spring into her arms and take up his accustomed position on the head of the household. He also learned tricks. His favorite involved a long straight wand that Mrs. Roberts used in the garden. Edward would climb halfway out along the wand, then scrunch up his body to bring his hands and feet together then—drumroll, please—he would twirl around at astonishing speed like an acrobat. Long after the age at which baby koalas are abandoned by their mothers in the wild, Edward continued to cling to Mrs. Roberts. He slept in a cot in her room, and he spent most of his waking moments in her arms. The attachment was mutual. She couldn't bear to be separated from him and brought him along with her, plus a generous supply of eucalyptus, on her travels to Europe and the mainland.

On major donors' night, a number of dignitaries came in for a cocktail party in front of the Carnivora House. Workmen had strung up a big blue ribbon across the entrance to K'Bluey's

exhibit. The zoo had wanted to get the mayor to cut the ribbon, but had to settle for Lucian Blackwell, a district council-
man who had been very helpful to the zoo. An Australian-style
band, in wide-brimmed Australian hats, was playing "Waltzing
Matilda," and there was a lot of Foster's Ale around to drink.

The keepers were supposed to be on display at these functions, and they milled around in their zoo clothes, acting uncomfortable. They were still feeling a little queasy from
yesterday's news that a young keeper named Robin Silverman
had been mauled to death by a Siberian tiger in the Bronx
Zoo's Wild Asia exhibit. They had found out from Dan Maloney, the young keeper at African Plains who had used to work
at the Bronx; a friend of his there had called him. Maloney
knew Silverman only slightly. The news spread rapidly around
the zoo. Apparently, Silverman had forgotten the cardinal rule
of zookeepers—always make sure any dangerous animals have
been taken out of the exhibit before you climb in—and had
walked into the tiger exhibit without realizing that the tigers
were there. Possibly she was distracted by a volunteer named
Barbara Burke who was at her side. Burke managed to get
away by scrambling up a sixteen-foot fence, but the tigers were
on Silverman in an instant.

Wilbur Amand and I talked about it over hors d'oeuvres.
Amand is a big, rolypoly fellow with, usually, a jolly sense of
humor. Having come to the zoo as a veterinarian, he was now
second-in-command to Bill Donaldson. Amand had run the
zoo for a period while the search committee did the work that
finally led to Donaldson's hiring. Amand himself had been a
candidate. When a committee member arrived somewhat anxiously to tell him that he hadn't been selected, Wilbur exclaimed, "Oh thank God! Let's have a drink!"

Amand is a member of the Mennonite sect, and looks it,
with a thick beard that runs under his chin. One grandfather
ran a dairy farm; the other had an agricultural farm where
Wilbur grew up. That farm used no tractors, only horses. Wilbur helped with the butchering from an early age.

Talking about the Silverman incident, Wilber grew sol-
emn. "It's hard to know what goes through an animal's mind
in a situation like that," he said, "whether it regards an in-
truder as food or a threat. Apparently, the tigers did not de-

vour the body, so it was probably the latter. They crushed her skull and broke her neck, and there were deep claw marks on her lower torso. The keepers got in and chased the animals away with fire extinguishers. I don't know if she died instantly or at the hospital. I've heard both."

"How do tigers kill a human?" I asked. It was a morbid question, but Wilbur kindly did not appear to mind.

"They generally do it by crushing your windpipe," he replied. "So you die by asphixiation."

He thought for a moment. "It's the same thing with lions. Bears claw you. They take you down with one swipe of their paws. They can break your back or snap your neck. Elephants like to skewer you with their tusks. Alligators drown you, unless you're small, then they crush you between their jaws."

"Which would you choose?" I asked.

"I'd prefer something quiet," he said. "Like euthanasia." He pondered this a little more. "You know, it's lucky that humans don't have to endure the kind of pain that animals do. I suppose war victims suffer a hard death, but it's not like animals. They don't die in their sleep. They're usually ripped apart. Some species, like antelope, have evolved a fairly merciful way to go. When they are grabbed, they go catatonic. There is an override somewhere, and they just go numb all over. I don't think they feel a thing."

We sampled some sliced fruit off a tray.

"You know, there was an incident a few years back at the Central Park Zoo where a guy thought he could talk to the animals. He was a professor. He'd tried to get a job and failed. He was feeling increasing distress. One night he climbed into the polar bear den and the bears mauled him. That was pretty ugly. There was another occasion back in 1958 when a tiger killed a youngster at the National Zoo. I remember that back then there were demands from the public that the tiger be killed. It's good to see that, with this latest incident, no one has called for euthanizing the tiger here." (Nearly two years after Amand spoke, police shot two polar bears named Teddy and Lucy at Brooklyn's Prospect Park Zoo in May 1987 after one of the bears had killed a twelve-year-old boy. Apparently on a dare, the youngster had sneaked into the bears' enclosure to wade in their pool after hours. He was set upon by the bears immediately. When the officers arrived, they were afraid that two other children might also be in danger and, with twenty blasts from 12-gauge shotguns, they killed the animals. "[The bears] were not executed," declared city parks commissioner Henry J. Stern. "The police did the right thing under the circumstances.")

The guests were all decked out in their party clothes and eager to see the koala. I tried to scout out some of the donors, but the only one that the keepers were able to identify was Kippie Palumbo. A lean figure with a theatrical look, she was wearing an unmistakably major donors' dinner dress of bold black and white, and a silver raccoon dangled from a gold chain around her neck. Kippie was the widow of Frank Palumbo, the owner of Palumbo's Restaurant in Philadelphia's Italian district, who in his passion for the zoo, had donated hundreds of animals, including the hippo who nearly beheaded his mate during sex. Somewhat confusingly, Palumbo had named him Frankie after his daughter, Franca. He also contributed a series of bald eagles, each of which he named Kippie.

"Everybody in the family was an animal," Kippie told me. "We all had animal names. My husband, Frank, was a polar bear. My daughter, Franca, she was a hippo because she was so slim. She only weighed a hundred pounds. That's why the hippo my husband gave to the zoo is named after her. All sorts of people started giving her little hippos. You know, ones carved out of wood, some silver ones, and some perfume ones. I mean, they were holding perfume. My son, Frank, Jr., he was a koala, because he was so cuddly and adorable. We even called him Koala when he was growing up. Have you seen the koala here?" I nodded. "Isn't he cute?" I nodded again. And she herself? I asked. Was she by any chance an eagle? "Oh no," she said. "I was an elf."

Then Kippie excused herself and disappeared into the crowd.

I found the man from American Airlines who had arranged to fly the koala and the eucalyptus to the zoo for free. "We carry animals pretty regularly, you know," he told me, "but usually they're pets, and they're not up in first class." He said that the airline was happy to help out; I asked him if he would have done the same for an alligator. "They are a bit frightening," he said, retreating from me.

The crowd quieted; Elaine Chu had brought out K'bluey. She carried him on her gloved arm as if he were a ventriloquist's dummy. Bill Donaldson showed that his years in city politics were not wasted as he praised the koala's many sponsors and introduced Lucian Blackwell, who represented the zoo's district. Blackwell snipped the ribbon and handed K'bluey a eucalyptus bouquet, complete with roses and baby's breath. Elaine accepted it for the koala. And as she did, K'bluey leaned forward and planted a kiss on Elaine's cheek. The crowd applauded. Then Roseann Giambro, the gorilla keeper, stepped up to give Elaine a key to the zoo, an oversized version of the "zoo key" that turned on various information tapes. Elaine glowed appreciatively, then fished the eucalyptus out of the bouquet, and handed it to K'bluey, who sniffed it, nibbled it, and scarfed it down. Elaine took him back to his cage. Not knowing what else to do, the crowd applauded again, then returned to the hors d'oeuvres as the band struck up "Tie Me Kangaroo Down, Sport."

—10—

Penrose Laboratory is a big stone building that rises up behind a gate marked NOT OPEN TO THE PUBLIC. It is named after Charles Penrose, president of the zoo from 1907 to 1925. Wealthy from a family shipping fortune, Charles Penrose was a professor at the medical school of the University of Pennsylvania, specializing in gynecology. In 1905 he founded the Penrose Research Laboratory, the first such animal laboratory in the world, in part to answer a question that had long puzzled him: why childbearing was so painful for humans but relatively painless for animals. As far as he could discover, the answer lay in the differences in pelvic structure between humans and animals, and in the different levels of awareness of what is happening during birth. As Dr. Snyder explains, "If a woman is dreading the birth ahead of time, she's going to tighten up the so-called smooth muscles of the uterus and it's more painful. For animals, it comes naturally."

Charles Penrose himself was also the brother of the famous Boies Penrose, who as a U.S. senator ran the Republican party in Pennsylvania and the nation like a Mafia capo. He put Teddy Roosevelt on the ticket as vice-president under McKinley. Boies ate eggs by the dozen and drank coffee by the quart. He was as rugged as his learned brother Charles was tame. There is a story that the two brothers once went out hunting for grizzlies in the Grand Tetons. When Charles got mauled by a wounded bear, Boies lashed him to his saddle and, leaving their guides behind, bushwhacked cross-country to the nearest railroad to take him to the hospital.

Penrose Laboratory is actually something of a misnomer, since it is only partially a lab. Most of it is a rabbit warren of offices—upstairs for Dr. Snyder, Wilbur Amand, and the three curators, and downstairs for senior keeper Bill Maloney and the veterinarians. Even though there were only a few small animals around, you could tell you were still at the zoo by the aroma, the animal posters, and by the number of Gary Larson "Far Side" cartoons tacked up on the bulletin boards.

Eileen Gallagher is a veterinary technician in the downstairs lab. It is the equivalent of a laboratory in a medical hospital. Eileen spends most of her time with her head down, poring over a slide on her microscope as she tries to get a fix on zoo life at a molecular level. She does hematology, urinalysis, microbiology. At this particular moment, she was counting blood cells from a blood sample of an ailing kangaroo.

Eileen is a pretty twenty-five-year-old who wears a little eye makeup that adds a touch of color to her Khaki zoo uniform. Her earrings and belt buckles usually bear animal figures. Today, it was leopards. She is always friendly. Since her lab is located right by the Penrose front door, it is a frequent stop for the keepers, any number of whom have a crush on her. For her birthday one year, the crew from the Elephant House tied a bow on a urine sample from Billy the rhino. For Christmas, they gave her some of what she calls Billy's "do-do." And a keeper from the Small Mammal House often urged her to come see his collection of feathers and whiskers. Eileen collects tiger toenails and tenrec teeth, but she always declined the invitation. Her passions run to animals, and stay there.

Eileen introduced me to Ralph, the binturong she was raising, who had assumed his accustomed position curled around her neck like a scarf. "He likes being up there so that he can see everything," she said. Ralph looked like an elongated kitten, but his black coat was rougher, and his tail longer and thicker. The tail is prehensile; he could dangle from it. "He lives with me and my dog," she said. "That's Roxanne. She's half German shepherd and half beagle. When I was growing up, our dogs were always named Spot or Whiskers. I wanted a really good name, so I named her after that song by the Police. She loves Ralph as much as I do." She gave Ralph a pat and a kiss, which went unacknowledged. "Roxanne was raised with a baboon, so she understands about wild animals. She and that baboon had a great relationship. She let the baboon ride around on her back. They were inseparable. When I introduced Roxanne to Ralph, I let her come and sniff him and everything. I told her that he was just a baby, so she had to be nice to him. Ralph rides on her back, and she doesn't mind. She doesn't even really mind when Ralph chews on her foot."

"Ralph seems to like you, too," I observed.

"Oh yeah. When I'm home, Ralph's always up on my shoulder, or he's in my lap. He sleeps in my bed with me. He lies on the bed like a cat, usually curled around my legs or in a ball on my stomach. It's all right. He only weighs three pounds. Binturongs are nocturnal, so he gets up at three in the morning, and he tries to get me up by licking the corner of my eye. It feels kinda scratchy. But I don't mind."

Eileen wasn't paid to raise animals, but she did it anyway. Fellow veterinary technician Ann Hess took most of the orphaned animals, but Ann had been plagued by injuries of late, most recently a nasty bite on her index finger that she received from a woodchuck over the winter. When the woodchuck jumped off a countertop, Ann instinctively reached out to keep it from hitting the floor. The woodchuck responded by sinking its teeth in Ann's finger. She had been out for several weeks, off and on, recuperating.

Although Eileen felt bad for Ann, she was certainly glad to have a chance to raise Ralph. Even though it meant giving up her weekends and being woken up every morning at three A.M., she regarded it as her reward for putting in so many hours over the microscope. She'd always liked animals. "There are pictures of me this high walking up to deer and bears and stuff," she said. And at last year's high-school reunion, she was the only person in her class who had never changed her mind about her career. Like many people at the zoo, Eileen had come to be with the animals, only to find that few zoo jobs there actually involved animals on any regular basis. So she tried to fit them in where she could. It was the same story with Beth Bahner, another strikingly attractive young woman who worked upstairs as a kind of general secretary for the lab. She was stuck behind her desk four days a week, and lived for the fifth, Friday, when she got to put on her laboratory coat and help the vets by holding down the animals as they strained on the operating table. Eileen did some of that, too. She held the clip-board when the vets went out on a knockdown (anesthetizing an animal in order to give it medical attention), and she re-corded the dosages, times, and all the other technical infor-mation. She tagged along with them; they treated her like their kid sister.

But Eileen also undertook the far more involving task of raising some of the animals. She had raised marmosets, a kan-garoo, a mandrill called Barbara Mandrill, a bristly pigmy hedgehog tenrec, and a fruit bat. She loved every one, but none as much as Ralph. With Ralph, the attachment was phys-ical—he was always on her somewhere. She dreamed of build-ing a spare room off the apartment she shared with a roommate for the animals. She didn't mind having them dash around the apartment, but she always worried that they were going to chew on an electrical cord. The spare room would be animal-proofed, the way some houses are baby-proofed. The animals were her babies, and she lavished on them a mother's love. But there was tragedy built into all of these relationships, since her task was to raise them, not to keep them. When they were grown, something that happened all too fast, they would return to their exhibits or, more likely, be sent off to another zoo. Either way, they would leave, and Eileen would be devastated. She tried not to think about it, but the end of her relationship with Ralph was drawing nearer every day.

The phone rang, pulling Eileen away, so I took my leave and went off to find the vets.

As usual, they weren't in their office. Of all the zoo staf-fers, the veterinarians were probably the most overworked. They had the most far-ranging responsibilities. For it was their task to keep the 1,700 animals in the collection alive and well;

sometimes it seemed to them that the zoo was nothing more than a 1,700-bed hospital. In a regular hospital, of course, all the patients were of the same much-studied species. While rudimentary information—average body weight, standard internal temperature—for most of the exotic species under the vets' care was complete, it hadn't come anywhere close to the level of detail available for humans. In diagnosing the extent of heart disease, for instance, it is useful to know any variation in serum enzyme levels. The standards are well established for humans; they are completely unknown for, say, a coatimundi.

Animals don't tell you how they are feeling either. Just the opposite. In the wild they have good reason to pretend everything is fine even when they are in the last stages of terminal illness. Predators show no sympathy. And this disposition carries over to the zoo. Consequently, the keepers are supposed to scrutinize their animals hourly, and alert the vets at the first sign that something is wrong—if the animals eat less than usual, if their stool is loose, if they are touchy, if they are favoring one hoof. A major part of the vets' job is to cultivate the keepers. A previous veterinarian performed this part of the job with astonishing insensitivity; it got so bad that some keepers deliberately withheld information from her in the belief that the animal was better off in their care than in hers. Animals they sent to her never came back; they went on to the morgue.

Keith Hinshaw and Mike Barrie suffered no such problems. The animals might cringe when they walked into the room, but the keepers breathed easier. They were confident that their complaints were taken seriously, and that the animal would soon be restored to health. If it wasn't, they could be sure that everything that could be done had been done.

Keith and Mike act like a couple of cops, the sort who always know what the other is thinking. Keith is the senior man, and the official zoo veterinarian. He had been here since 1982. Mike was here for two years on a government grant to learn exotic veterinary medicine and then push on to another zoo. His term would be up next July, and neither of them was looking forward to it. Like cops, they both wear bulky walkie-talkies on their belts; they also wear khaki uniforms—shorts in hot weather—and work boots. Keith is tall and almost unnat-

SO. ST. PAUL PUBLIC LIBRARY
106 3RD AVE. N.
SO. ST. PAUL, MN 55075

urally slim. He has craggy features and wears a reddish beard that would not look out of place on a logger. Women find him very attractive, but he has yet to marry. Mike has a fuller physique but thinner hair. He is married to a fellow veterinarian named Kathy who specializes in small-animal medicine, usually dogs and cats.

"Don't step on the patient," Mike sang out cheerily as I strolled into the Penrose Annex next door that doubled as the zoo hospital. I looked down. A very sickly woodchuck was lying at my feet with its tongue out, waiting its turn on the operating table. Right now, a kangaroo was up there, sprawled out on its side under the big lamps, its long tail dangling down over the edge. Keith and Mike stood over it, puzzled.

They were acting jolly, but they were distressed by what they saw. Keith had nearly killed himself over this animal, whose name was Blasee. He had broken his leg a few months back, and with enormous difficulty Keith managed to place a pin in it. To fight off infection, he had to put Blasee on antibiotics, and Keith took it upon himself to inject the kangaroo with them three times a day—at eight A.M., four P.M., and midnight, even though that meant leaving his house at eleven, when he would normally be heading for bed, and driving a half hour to the zoo from his house in Manayunk, a section of Philadelphia. Then he faced a tedious, rather risky procedure (especially in the dark) of climbing into the kangaroo enclosure, grabbing Blasee by the tail, and pulling the animal to him. "One night," he told me, "it was really dark, and I really wanted to stay home. I was exhausted. But I drive in, get the drugs, and go into his pen. I grab his tail, and I've got it between my legs when my hands slip and the tail goes straight up into my groin and sends me flying into the air. That really hurts. Then I come down right on his water bowl. It flips over and drenches me. But I give him his shot. Then I hobble back to my car, all hunched over. I finally got the pin out a few weeks ago, and now look at him. After all that, I want him to live."

Blasee was literally dying of diarrhea. His body was drying out and weakening. His kidneys were closing down. Keith had a stethoscope to check the kangaroo's heart. Mike had inserted a probe up the kangaroo's anus to check his temperature. "It's one hundred point six," he told Keith.

"What's normal?" I asked.

"Hundred one, hundred two, somewhere in there."

"How do you know that?"

"I don't. I looked in the book."

"So how's he doing?"

"Well, he isn't going to stand up and wave."

They ran more tests; some go to Eileen, some to a local hospital, Philadelphia Presbyterian, for analysis. They call Presbyterian a "human hospital" to distinguish it from the veterinary kind. Their lab work is done free during off hours in exchange for zoo passes. The hospital technicians who do the work sometimes don't know that the patient they are working on is not human, and express some surprise about the numerical values on their instruments.

The vets carried the fifty-pound animal next door and put him under a heat lamp. Today, they were war doctors, trying to make the dying as comfortable as possible. This was the third kangaroo from the enclosure this week to come down with the illness. They didn't know for sure what it was, but they had a good idea. It could well be toxoplasmosis, an insidious parasite carried in by stray cats, sometimes mice. This is one reason the zoo is so strict about not allowing house pets inside the gates, and it tries hard to secure its fences. Keith himself had seen a cat prowling in the kangaroo exhibit. Once a stray is in the garden, it can bring down any number of animals. It carries the parasite in its intestines, and spreads it in its droppings, which can easily be ingested by a zoo animal as it grazes. Sometimes the vapors from the excrement can transmit the disease.

The vets had to leave Blasee under the lamps in the adjoining infirmary for a while. The best they could do was try to make him comfortable and hope that some Kaopectate would help the diarrhea. They had other animals to attend to out in the garden—starting with a tahr that had some horn trouble.

We went out to the veterinarians' blue van that was always parked by the front door—except when a visitor took the space. Keith carried his E-bag, which looked like the sort a brush salesman might use for samples. The E is for emergency, but the vets don't take that literally. For real emergencies, they bring out the K-bag. That's K for Klyde, a veterinarian at the University of Pennsylvania who had provided them with powerful stimulants to restart an animal's heart and lungs. The M-bag is for general medicine.

Keith took the wheel and we headed off. Unlike some other zoos, the Philadelphia has no private staff roads, which meant Keith had to inch down the walkways while the visitors scattered in front of us. We passed by Solitude, then the administration building, to the African Plains. At the exhibit, senior keeper Bill Maloney was already waiting for us, along with his assistant, Dave Wood, and Ray Hance, a particularly agile keeper who was good at catching loose animals. These keepers needed to be good, because tahrs, mountain goats that are used to prancing about in the Himalayas, are jumpy. There are some six-foot cement pillars in the exhibit—they're for the tahrs to hop up onto. The goat in question was the ranking male, a great bull of an animal with an impressively shaggy chest and a stern expression. Maloney, Wood, and Hance needed to narrow the tahr's range by luring the animal into a smaller holding area in back of the exhibit. Like ranch hands in a corral, they stalked the animal with their arms outstretched, fanning the tahr toward the gate into the holding area. The tahr responded by galloping around the enclosure like a filly. Finally, he was cornered, and he retreated into the holding area, thinking that he'd done something clever. There, Dave and the tahr faced off about five feet apart, as if they were contemplating butting heads. Suddenly Dave sprang forward, and just managed to grab the tahr's horns before he dropped into the mud. We were a little surprised, and the tahr was bamboozled. "C'mon guys, help!" Dave screamed, and the other keepers closed in to hold the tahr down.

"He's pretty docile today," said Keith.

I looked at Keith skeptically.

"He charged at us today."

The tahr looked as though he were plotting something, and I said so to Keith.

"He's not going to do anything," said Keith. "They let you know if they're mad. They signal their aggression by looking at you. It's aggressive to look you right in the eye. Watch out if he looks like he's smiling." The tahr had his head down. I crouched down to check his expression. Keith laughed. I felt foolish.

Mike brought out his camera for a picture. "Now, smile everybody." Everyone groaned. "We photograph everything that we haven't seen before," he told me.

"You go through many pictures?"

"About one roll a week."

Everyone had a hand on the tahr's horn, as if it were a baseball bat and we were choosing up sides. The horn was long and beautifully curved, like a musical instrument. But at the base, where the horns were rooted to the animal's head, a cut had opened up, some insects had gotten inside and had started to breed. The base of the horn looked like a beehive. It was dark and ugly with pus and larvae. It was possible the holes went clear down to his sinuses, turning the horn into a third nostril. The tahr stood stock-still, his eyes wide. Keith pulled out a long Q-Tip and swabbed out the holes. Then he squirted in some disinfectant with a syringe, and finally slathered the whole thing with Equi-shield fly repellent to keep the bugs off. "All you can do with something like this is try to keep it clean," he said. "And you hope that the bone hasn't rotted out underneath. If the bone goes, then you've got to cut it out." Keith didn't relish the idea of scraping away at the animal's skull. He wouldn't want to have to replace the horn either. The zoo dentist, Dr. Carl Tinkelman, could do it with some dental acrylic. He had replaced a blesbok horn a few years back, but that sort of thing almost never came out right.

When Keith was done, the keepers released the tahr, and he dashed off and ran around and around in his pen deliriously before returning to his family. The vets returned to their clipboard. They had to go on to examine some strange nodules that had sprouted on the head of Salty, the sea lion down at the Children's Zoo; weigh a baby trumpeter swan that was wasting away; and, if there was time, check some new flooring developed for aircraft-carrier runways that the zoo was trying out for the giraffes. If a floor is too slippery, the giraffes do the splits; if it's too scratchy, they get terrible blisters. This might be just the right point in between.

Then the vets had to get back to work on the woodchuck they left on the operating-room doorstep, and to look in once more on Blasee the kangaroo. It figured to be a long day.

If they were lucky, they might have a chance to take another peek at K'bluey. Until K'bluey came in a few days ago, they'd never seen a koala before.

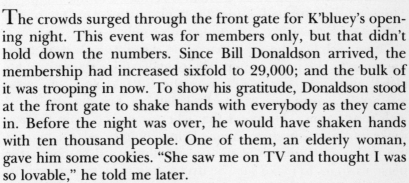

The crowds surged through the front gate for K'bluey's opening night. This event was for members only, but that didn't hold down the numbers. Since Bill Donaldson arrived, the membership had increased sixfold to 29,000; and the bulk of it was trooping in now. To show his gratitude, Donaldson stood at the front gate to shake hands with everybody as they came in. Before the night was over, he would have shaken hands with ten thousand people. One of them, an elderly woman, gave him some cookies. "She saw me on TV and thought I was so lovable," he told me later.

Unlike the usual zoo visitors, who never seem to know where they are going, these folks charged ahead purposefully. The parents tugged children, and the children held tight to stuffed koala bears and other toy animals they'd brought along like mitts to a ball park. Of course, these visitors really didn't know where they were going either; a host of volunteers, bedecked in ribbons, had been arrayed to point the way to the Carnivora House. "This way for the koala!" they yelled. "This way for the koala!" There, ropes channeled the crowd around a sculpture of a lioness feeding a boar to her cubs, and then into the exhibit. The line grew. To keep everybody happy while they waited, the zoo brought out jugglers, mimes, more Aus-

tralian musicians, face painters (to apply koala noses to youngsters), a videotape showing the making of the Quantas
commercial (that was a big hit—some people saw the koala,
then got back in line just to see the Quantas tape again), some
jars of eucalyptus to smell, a woman cavorting about in a koala
costume, and a crew of docents who circulated through the
crowd explaining over and over that the koala is not a bear
but a marsupial.

Before long, the line wound clear back to the elephant
house, and it took over an hour—at the end of which the visitors got to see K'bluey for about four seconds. This evening,
as usual, he was asleep. He was hidden behind the foliage at
the top of his tree. If the visitors strained, they could make out
a patch of ash-gray fur through the leaves. They might even
have gotten a sense of K'bluey's general shape. I learned later
that he stirred once and, according to some observers, popped
his head up and opened his eyes a crack. The viewers cheered.
Then K'bluey dropped his head and returned to his slumbers.
The visitors continued to file slowly past as before. One of the
staffers said it was like visiting Lenin's tomb.

Scott Schultz, the zoo's marketing director, looked on from
the fence by the tiger exhibit. He is small, with a dapper moustache, and came to the job from the Buffalo Zoo, where he
was assistant director. At the Philadelphia, Schultz also helps
with fund-raising. He notes that when someone from the zoo
walks into a prospective donor's living room to ask for money,
the donor immediately casts an eye downward to check the
person's shoes to make sure he isn't tracking anything unpleasant onto the Original rugs. Schultz relies on Bill Donaldson to
cement any agreement, which Bill often does in what he
cheerfully terms his "fleecing tent"—an evening barbecue he
hosts at the zoo for prospective donors.

More than anyone else, Schultz was responsible for the
increase in membership, and for all the publicity. Nevertheless, he watched the gathering crowd in mounting agony. "I
hate lines," he said. "This just kills me." The problem was that
while last year's white tigers could be exhibited on the rocky,
moated island behind him, and be seen by a thousand people
at once, the koala was so small he had to be viewed individually, and that took forever.

Still, Schultz knew the zoo needed the koala to dispel a faint air of ennui that hung about the place. Loyal as Philadelphians may feel toward their zoo, they can be a little sluggish about turning out. For them, zoo visits are, as he says, "infinitely postponable." Why go today when they can go tomorrow? Or next year? More alarming, visitors are inclined to forget how long it has been since they last went. They think it's been a year, when it in fact has been two or three. An exciting, heavily publicized, one-time event like the coming of the koala breaks the cycle. Then they *have* to go today, or they miss out.

So the key to the success of the koala's visit was brevity. In fact, even if the koala could stay forever, the zoo might very well be tempted to send him back after a month or two. Or, more deviously, the zoo might *claim* that he was here for a limited time only. The administration had done that with a snow leopard a couple of years back. When the leopard had arrived, they had declared with great fanfare that the animal was here for two months only. When the two months had elapsed, they said that the animal had been held over by popular demand. The snow leopard was still here. He occupied a cage in the Carnivora House. Viewers walked right by him to see K'bluey, but they didn't notice.

Kathy Wagner, another vice-president, strode up with a tiny koala pin clinging to the neck of her dress. Donaldson walked along with her. He had taken a respite from handshaking to see how the lines were shaping up.

"Well, we got our crowd," Kathy said.

"We sure did," said Scott.

"But K'bluey is asleep," she said. "He's missing the whole thing."

"How about a little electrical current?" said Schultz's assistant, Steve VanderMeer. "Every time he nods off, we zap him."

The group fell silent. They were too grateful to K'bluey to make fun of him—yet.

"Well," said Donaldson, "back I go."

Scott watched him depart. "If you want to know why this zoo has been successful, there's your answer." We looked at Donaldson's retreating form. "Bill's the whole thing. His personality is so completely associated with our zoo. He is *so* important to our success. You can't show it on paper. But he has done everything for us. He is the Philadelphia Zoo. He is warm, funny, charming."

"And he isn't on loan from the San Diego Zoo," I said.

That stopped Scott for a second. "Right," he said with a smile. "And he isn't on loan from the San Diego Zoo."

—12—

In the Penrose Annex the next morning, a wallaby—a smaller version of a kangaroo—was on the operating table. The vets had first heard he was sick yesterday from a keeper who told them there was a wallaby that was having trouble standing up. They had brought him in from the enclosure the wallabies share with the kangaroos and some Aleutian geese, who seemed quite interested in the proceedings. They'd weighed him, and taken a blood sample, then had put him up in a special shed near the wallaby yard that they use as a sick bay. It looked like another case of toxoplasmosis.

This morning, the animal was worse. There was a nasty brown stain under his tail, which flopped over the side of the table. "He's got a bad case of the squirts," Mike Barrie told me. Mike was probing the wallaby's nether region with a gloved hand trying to see where to insert a catheter.

"Can't find his wee-wee?" asked Eileen. She had a hippo pin on her shirt this morning. Beth Bahner was helping out today as well, since it was Friday.

"I know it's here somewhere."

"*Good!*" exclaimed Dave Wood. He was wearing a furniture mover's jumpsuit. He had stretched his body across the wallaby's shoulders to hold him down in a passionate embrace. He and the wallaby were practically nose to nose. The wallaby's eyes were wide, as if stunned. Keith had told me that animals go numb when they are about to die. The wallaby seemed numb now. Dave held him down tightly all the same.

"Why is it that I can go all day without scratching my nose once," Dave grumbled, "but then as soon as I have to hold an animal down, I get this unbelievable itch?"

Eileen scratched him.

"Oh, hey, that was nice."

"Here we go," said Mike, and he attached the plastic catheter.

The wallaby squirmed feebly.

"You gotta have a lot more fight than that," said Dave.

Mike shaved the inside of the rear leg, then took out a needle to draw some blood.

"Is he under anesthesia?" I asked.

"He's not under anesthesia," said Mike. "He's under illness. Besides, animals don't respond to surgical procedures as much as humans. They run around with major pain. You spay a dog, and he'll hop right off the table as though nothing happened. But this guy is past caring." He plunged the needle in.

"Okay. I got the vein," he told Keith.

There was silence for a moment.

"But there's no blood in it."

"Try another one."

He tried another one.

"There's no blood in there either," said Mike.

"He doesn't want to give any up," said Keith, sounding low-key as usual. "I don't blame him. Try the jugular. Didn't we try that on Blasee? That ought to have some."

"What are you going to do?" I asked.

"We're going to give him some fluids, some sterile electrolytes, and we're going to pray."

If it was toxoplasmosis, this would be the fourth animal from the kangaroo yard to come down with it. The first two, a kangaroo and a wallaby, were dead. The third animal, Blasee, was on his way out. He was still under the heat lamp in the next room, growing thinner and more feeble by the day. But his leg was fine. And now this one.

They took his weight. Mike stood on some bathroom scales, then Dave handed him the wallaby.

"These scales are suboptimal," Keith admitted.

"He means they suck," said Mike.

By subtracting Mike's weight from the total, they figured out that the wallaby weighed twenty-three pounds.

"Twenty-*three?*" asked Keith, not sure he heard right.

"Yeah, twenty-three."

"Sheesh, he's lost seven pounds just since yesterday."

They laid the wallaby back on the table and tried to force some peanut butter into his mouth with a tongue depressor.

"I'd like to get that damn cat," said Dave. "How do you eat cat, anyway?"

"I like mine with hot sauce," said Keith dryly.

"You know, the only thing that eats wallabies out in the wild are dingoes," said Mike.

"Them and thylacines," said Dave, referring to the strange pouched doglike creatures also known as Tasmanian wolves.

"But they're extinct," said Keith.

"They think there may be some left," said Mike.

"That would be great," said Dave. "The Australian guy who shot the last one—if it was the last one—he had his picture taken with it, he was so proud. It's like the great auk. You know—the bird. Some fishermen found the last two great auks on an island off Iceland. They were sitting on a nest, and the nest had an egg in it. What did the fishermen do? They clubbed the two auks and then smashed the egg against the rocks. Can you imagine? It's the same mentality as the weirdos who shoot at the president.

"So that's why we want to save you, little guy," said Dave, looking deep into the frightened eyes of the wallaby.

"Don't kiss him," said Eileen. "He's got enough diseases."

Keith read off the results of the bloc analysis. "His glucose level is off the scale. We better put in some dextrose to counteract."

They inserted a speculum in his mouth to force the liquid down.

"Should we give him some more peanut butter?"

"He didn't really go for it," said Keith.

"That's because you didn't give him any jelly," said Beth. "Let's stick to Kaopectate."

"That's the yellowish stuff?" I asked.

"Yeah, they load it up with food coloring to make you think it's doing something. It's funny, they always color things more brightly in animal medicine than in human medicine. Like these bandages." He handed Mike some bright red bandages for the wallaby's wounds. "Humans get beige and earth tones. Animals get green, gold, red, and blue."

"The trouble is," said Dave, "you put an animal back in the pen with one of these colored bandages, and they all want one."

Keith shone a pen light into the wallaby's mouth, checking for dental problems. Mike got a syringe and thrust it into the wallaby's abdomen to get a urine sample to send to the University of Pennsylvania vet-school labs.

"We'll give him an IV to replace the fluids he's losing," said Keith. "It sure looks like the toxo, but we'll hit him with a broad spectrum of antibiotics. We'll try to cover as much as possible."

"Any chance for the wallaby?" I ask.

"Well, we're zero for two so far," he replies. "The good news is that some of his signs are different from the other animals that died of toxo. The bad news is that the Wallaby is real sick."

Dave Wood put the wallaby's paws together as though he were praying.

The three men carried the animal back to his sick bay. Eileen followed, carrying the wallaby's tail. They set him down gently in the corner of the shed. He hardly moved; only his belly swelled irregularly as the air pumped through. The others drifted back to the lab, but Keith and I stayed behind for a minute. The wallaby was curled up in the corner, sagging like a deflated football. Not even his powerful tail had any life in it. It hung as limp as a wind sock on a calm day. Keith looked at the animal sadly and shook his head.

"Hurts you, doesn't it?" I said.

"Yep," he said. "Sure does."

——— 13 ———

At the middle of August, Rick Biddle was in despair. Rain had wiped out some key weekends, and scorching heat had kept the crowds down much of the rest of the time. Biddle's hopes for the elusive 300,000 had wilted. Even the estimate of 250,000 seemed too rosy. It looked to Biddle as though the zoo might well come in at 200,000. "At this point," he said bleakly, "I'd be real happy with anything above two hundred ten." Tolerable for August, but nothing to build on. He'd already made tentative plans to scale back the development projects to reflect the disappointing outcome.

Then later in the month, his luck turned. The zoo had planned a special birthday party, co-sponsored by local radio station WDAS, for K'bluey on Saturday, August 24. The event fell on a glorious day of bright sunshine and gentle breezes, and the crowds poured in.

That turned things around. The weather stayed pleasant, and people flocked in to see the koala. They came from the inner city, from the Jersey shore—thus justifying the banners the zoo paid to have flown behind airplanes in New Jersey—and from as far away as Canada, providing a finishing kick that pushed the total up to 251,000.

Rick and his wife, Ann, came in on Labor Day weekend for K'bluey's last day.

"Did you give him a hug?" I asked.

"To tell you the truth, I didn't," he said. "But right before they took him off exhibit, I went in and told him good-bye. I also said, 'Thanks.' And Ann and I waved to him as he left."

The next morning, Monday, September 3, as soon as the final figures came in, Biddle sat down at his desk to write out the order to restore the plans for the picnic grove, the redesign of the Bird house, and for new lighting in the garden. However, because the full 300,000 hadn't come through, Biddle would have to hold off on the animal holding area he wanted to build. But that was a small price to pay to avoid catastrophe. "I'm glad we came out of the month on course," he says. "August can be a bitch. If it's not broiling hot, it's pouring rain. Now I can relax a little bit."

He didn't look relaxed in the least.

Talking about the koala's visit afterward, staffers were struck by how quiet everybody was when they finally got in to see the koala. Scott Schultz said it was like a "religious ceremony." Someone else suggested that the visitors might not have wanted to wake K'bluey up. And another staffer joked that the zoo should immediately advertise K'bluey's return, and then position a stuffed koala in the tree. "Nobody would notice," he said. And the presents were still pouring in. The prize for the most unusual gift had to go to the clip-on fingernail bearing K'bluey's portrait. Elaine Chu got that and like a good sport, she hung it around her neck on a velvet ribbon. "K'bluey came, he saw, and he conquered," concluded Beth Bahner.

K'bluey also profited from the experience. He gained several pounds—always a good sign of a young animal's well-being. And he learned how to produce that "cry of the koala" that Ambrose Pratt wrote about which is the mark of the mature male. He was sitting in the ladies' changing room with Elaine. Wilbur Amand had popped his head in for a visit, when suddenly K'bluey reared his head back and seemed to be struggling for breath. "I thought he was having an epileptic seizure," Wilbur called. And then out came a piercing shriek: "A-ah-aaaaaaaaaaaaaahhhhhhh!" He looked immensely satisfied for a moment. Then, little by little, he drifted back to sleep.

K'bluey had grown up.

—14—

For Eileen Gallagher, the day she had been dreading all summer came September 3. A couple of young staffers from the Bronx Zoo named Jimmy and Ruth arrived in a Datsun sedan to take Ralph, Eileen's binturong, to their Children's Zoo. Eileen had been steeling herself for this moment, but it was no use. The tears brimmed at the corners of her eyes, smudging her mascara. The people from the Bronx looked uncomfortable. "I feel like I'm from the orphanage, coming to take your kid," Jimmy said.

"Oh, don't worry about it."

Eileen had sliced up some apples and bananas for the ride. "Ralph tends to cry in his crate," she told them. "These might help."

"Okay."

"And he kind of likes this." She put Ralph's teddy bear in the crate. It was a little bigger than Ralph, and tan-colored with dark ears. It looked, in fact, a bit like a koala. It was a little ratty in spots where Ralph had pulled at it overzealously with his emerging claws. Theoretically, the stuffed animal was zoo property, but Eileen was not about to stick to the rules now.

"Good idea," Jimmy said.

Eileen had to fill out an Animal Transfer Form, so she handed Ralph to me for a moment. He grabbed on to my arm, scooted up to my shoulder, and huddled there. He was as clingy and lightfooted as a kitten, but his whiskers were longer and his black fur, flecked with gray, rougher. I felt honored to have him on me.

On the form, below the spaces for weight, age, and animal identification number, Eileen tried to give a description of Ralph's character. "Ralph is very playful and loving," she wrote in a wobbly script. "He's used to a lot of attention. He'll follow you everywhere. His favorite place to be is on someone's shoulder." With that, she ran out of space, although she had a lot more to say.

Ralph climbed back aboard Eileen when she finished writing. She had to loosen his claws from her work shirt to get him into his traveling crate. But once inside, he appeared not to mind, just paced around uncertainly. She carried the crate out to the car and with the Bronx staffers looking on, put it on the backseat. Despite efforts to hold them back, the tears spilled down her face. She wiped them away with a free hand.

"Take care of him," she said.

"Don't you worry," the man replied.

The two staffers climbed into the car and started the engine.

"You wouldn't mind if I came to visit him, would you?" Eileen asked timidly.

" 'Course not. Come any time."

"Okay. Well, see ya."

" 'Bye."

She looked in the back window. " 'Bye, Ralph."

The car went off. Eileen and I watched till it turned out of sight, then we went back into Penrose. She returned to her microscope, but she couldn't concentrate. Her mind was still on Ralph.

"He's up to four pounds," she said.

"I saw on the form," I said.

"He's gained a pound. You can feel it. He's getting all grown up. It was time for him to go, I realize that. It was Dietrich's decision." Dietrich Schaaf was the curator of mammals. "You can get money for a young binturong, but not for an adult. They had to sell him now if they were going to, and there isn't space for him here. There's a market for the young ones because they're still trainable for education and PR things. They're easier to handle. They're like dogs. That's why Ralph is going to the Children's Zoo. They only bought him for one hundred dollars. I was going to buy him myself, if I could have had him de-scented." Mature binturongs' anal glands give off a strong odor that smells like burned popcorn; civet scent has been a staple of the perfume industry since the days of King Solomon. "But they don't know how to de-scent them yet. I'm going to have a hard time getting Ralph out of my mind. I worry that the people at the Bronx won't know how to take care of him. They won't know that he likes to play, to lie on his back and have his belly scratched, that he likes his teddy bear. I'll go see him at the end of the month. I'm going to see his sister Alice next week. She's at the Cincinnati Zoo, and I'm going there for a conference. Whenever I go to another zoo, I always go to see our animals.

"I'm sorry I won't get to watch Ralph grow up anymore. He was starting to act more like a binturong. He was really running around and climbing on things. He was hanging off the curtain, and running up the lamps. He'd climb and leap. He was really getting into leaping. That's a real binturong thing. He'd run up the banister, then fly over the couch and onto the coffee table. I don't mind curtains, but I do worry when he starts shimmying up the electrical cords. I trained him to walk on a leash. Seeing me, people would wonder why I had a kitten on a leash. We'd go everywhere together. And sometimes he'd jump onto the steering wheel when I was driving. I remember one time, he climbed on when I was backing up. I was looking behind me, so I didn't notice. I started spinning the wheel really fast to turn the car around, and he had to race to stay on like he was on an exercise wheel." She paused to savor the memory. "I'm going to miss him." She started to sniffle. "And Roxanne is really going to miss him, too."

The phone rang. "If that's Beth Ann, my roommate, I'm going to break down. She saw us so much together I can't talk to her without thinking of Ralph."

She picked up the receiver.

"Hello? Oh hi, Beth Ann. Where's Ralph? I guess he's on the expressway."

Eileen started to dissolve.

Outside her door, I saw Keith Hinshaw coming out of the pathology laboratory. He looked like a cross between a radio-activity cleanup-crew man and a zombie from *The Night of the Living Dead.* He wore a blue hospital gown, with a gauzelike cloth over most of his face, thick rubber gloves, and heavy boots. The cuffs of his pants were soaked in blood. He and Dr. Snyder had spent the day cutting up a dead eland in search of the pus-filled abscesses that indicate tuberculosis. Eland are handsome antelope. They have cowlike bodies and long, aristocratic heads with a pair of corkscrew horns that spiral up into the air. They once roamed all of Africa in prodigious numbers, but are now found only in African game preserves. One of the zoo's four specimens had died mysteriously after a period of frightening weight loss. Dr. Snyder's autopsy revealed a lesion on its spleen. That sent up red flags all around the zoo. Even Clarence "Binky" Wurtz, the chairman of the board, who normally stears clear of the day-to-day operation of the zoo, had gotten involved, calling in regularly to find out the latest developments.

Tuberculosis, the "white plague," is contagious—not only to other animals but to humans. It wasn't good PR. It put the toxoplasmosis episode into shadow: That disease had killed a couple of wallabies and kangaroos, including, finally, the two that Keith and Mike had tried so hard to save. If the TB got completely out of hand, it could wipe out the whole collection. The zoo had last had an outbreak twelve years ago when a young gorilla was brought in with the disease. Before they knew what had happened, the gorilla and two chimps were dead.

At the first sign of the tuberculosis, the zoo euthanized the three remaining eland, killing one a day for three days. A key factor in this decision was their monetary value. Rare as eland are becoming in the wild, they normally fetch only a few hundred dollars on the open market, not enough to cover moving costs, and they hadn't been moving at all of late. They were, in short, expendable; so down they went. The vets fought the disease like a forest fire—deliberately sacrificing certain animals in order to save others. And they began zoowide testing of susceptible animals, starting with the tahrs and the gibbons. This meant a long, exhausting process of catching each animal, then using a patch of loose skin, usually the eyelid, sometimes the folds under the tail, to run a test similar to the pinprick tine test used on humans. If the skin started to redden and swell in a few days, that was not good news. So far, all the animals were clean, but there were many more animals to go.

What made the event so harrowing was the report from an Iowa lab, to which the zoo had been sending blood samples of its four elephants to study their ovulation cycles, that the Asiatic elephant Peggy had tuberculosis, too. An eland was one thing; an elephant was another. It wasn't purely her monetary value, actually, that made the difference. While Peggy might have been worth $25,000 in her youth, that assessment had dropped to a mere $5,000 or $6,000 as she approached forty. But she was a celebrity to generations of zoogoers, so it was unthinkable to euthanize her except under the most dire circumstances. Nevertheless, to avoid contaminating the other three elephants, the zoo felt that it should isolate Peggy from them. The keepers were instructed to leave Peggy by herself in a separate section of the elephant house. The other three, all females, weren't having any of that. They believed in solidarity. They were so protective of Peggy's fellow Asiatic, Dulary, that when she lay down in the yard for her afternoon nap (an unusual thing right there, since elephants generally sleep on their feet), the other three always stood guard over her, shoulder to shoulder, facing outward like burly football linemen protecting a fallen quarterback. And they would keep that position until Dulary woke up an hour later. Now, when the other three discovered that Peggy was being separated from them, they went wild. It reminded the keepers of the day when the monorail train had run backward. All three elephants raced around and around the yard, shrieking hysterically and flapping their ears. The keepers had no alternative except to bring them back inside with Peggy. They weren't allowed in the same room with her, but they could see her through the bars of an adjoining enclosure. That put their minds at rest.

In the days that followed, the administration's hysteria died down as well. It turned out that none of the other eland had

TB. None of the other species had it either, although for safety's sake the vets continued to go methodically around the zoo to run their tests. Also, word came back from Iowa that the initial reading of Peggy's tuberculosis was in error. They were sorry about any inconvenience.

During the tuberculosis scare, a kangaroo abandoned one of her young from her pouch. The keeper found a tiny, scrawny character dragging himself across the kangaroo enclosure one morning. Because the infant was in serious medical condition, veterinarian Mike Barrie took charge of his rearing. But in a week or two, if the animal came along nicely, Mike said that Eileen Gallagher might have a hand in raising him. Eileen brightened noticeably. Mike insisted that it was too soon to start thinking of names because it wasn't clear that the animal would survive. Kangaroo mothers rarely reject their young without good reason, even if that reason is not apparent to humans. But Eileen was not going to be stopped. She was starting to think about names.

The lowland gorilla Massa, longtime Philadelphia Zoo celebrity, at forty-eight. By far the oldest gorilla in captivity, he died the night of his fifty-fourth birthday. *The Zoological Society of Philadelphia*

Bill Donaldson, zoo president, with a favorite squeeze—a boa constrictor from the zoo's collection *The Zoological Society of Philadelphia*

The neoclassical Bird House, built in 1914, with ducks and geese paddling around Bird Lake *The Zoological Society of Philadelphia*

K'bluey the koala sitting in a eucalyptus tree for a publicity shot. Usually he is asleep. *The Zoological Society of San Diego*

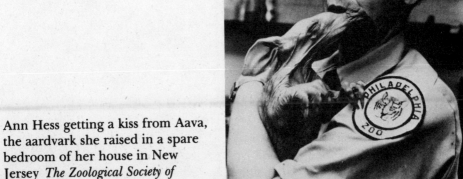

Ann Hess getting a kiss from Aava, the aardvark she raised in a spare bedroom of her house in New Jersey *The Zoological Society of Philadelphia*

Eileen Gallagher having a chat with Ralph, her beloved binturong *The Zoological Society of Philadelphia*

Keith Hinshaw, senior veterinarian, standing in front of the new primate center in an unusually relaxed moment *John Sedgwick*

The African Plains exhibit, with zebras and elands. The elands had to be euthanized shortly afterward because of the tuberculosis scare. This is a typical naturalistic exhibit that owes a lot to Carl Hagenbeck. Naturalistic, however, doesn't mean natural: The boulders and the baobab tree in the center are made of Gunite. *The Zoological Society of Philadelphia*

A kangaroo family.
F. Williamson/The Zoological Society of Philadelphia

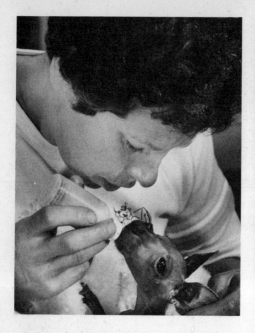

Ann Hess giving Eddie the kangaroo his morning formula *The Zoological Society of Philadelphia*

Boris, late brother of the slow loris Cloris. Lorises are prosimians from Southeast Asia. Their eyes are large so they can see in the dark. *Pat Kowal/ The Zoological Society of Philadelphia*

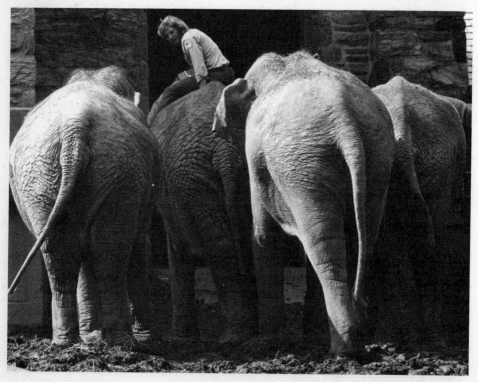

Ride 'em cowboy! Dave Wood sits atop Petal, leading the zoo's elephant pack into the Pachyderm House for the night. *Randy Ciurilino/The Zoological Society of Philadelphia*

Petal, on the right, and Kutenga do some friendly sparring with their tusks while Dulary minds her own business. *The Zoological Society of Philadelphia*

Petal and Kutenga saying hello to Xavira the Indian rhino *The Zoological Society of Philadelphia*

Dave Wood giving Xavira a pat *The Zoological Society of Philadelphia*

Fall

In the middle of September, a canvas L. L. Bean bag was hanging on the wall next to Eileen's desk in Penrose. It was the sort of bag you might put your towel and suntan lotion into when you went to the beach. From the way it sagged, I could tell there was something in it, and I asked Eileen what it was.

"Oh, that's Eddie," Eileen said casually without looking up from her microscope. "The kangaroo."

I thought she was joking. I pried apart the folds of the canvas bag and peered into the shadows. Inside, there were several rolls of imitation sheepskin wrapped around a grayish lump that was scrunched down deep in the bottom of the bag. A couple of rabbitlike ears poked up at odd angles. It really *was* a kangaroo. The animal's hair had only started to come in, so the top of his head still had that plucked-chicken look of the newborn. From what I could see of it, his body was crooked and flimsy. If Eddie's mother hadn't evicted him, he would still have been in her pouch, safe from my inquiring glances. The Bean bag was providing the next best thing: a soft, warm haven in a rough and chilly world. To keep his skin from drying out, Eileen rubbed it regularly with Vaseline.

"He's always looking for a pouch," Eileen said. "We let

him out to walk around. If I'm sitting down he comes up to me and tries to jump into my lap. Or he tries to jump into Barrie's shirt. That's the funniest thing. He dives at Mike's stomach headfirst. But he's taken the bag as his pouch. Whenever he sees it, he plunges right in."

It was good to hear Eileen sounding cheery again.

Much as she might like him to be, Eddie was not hers alone, though. Officially, the kangaroo was Mike's responsibility. He had been on duty when the animal was first discovered. He had tried to be careful with his emotions and not get too involved, because he knew that abandoned kangaroos rarely survive. But Eddie had come on so well in the last few weeks that Mike had let down his guard. He himself had named the kangaroo in the end. He said he chose "Eddie" because it was short and easy to pronounce. Eileen, disappointed, preferred to call Eddie "Spaghetti." To celebrate the birth of Mike and Kathy's first child that month, the couple sent out a photograph of the expanded family: Kathy holds the new baby; Mike holds Eddie.

Because of the tight bond that Mike had formed with Eddie, he usually took the kangaroo home with him, but occasionally Eileen got to baby-sit. And she often took care of him during the day, since Mike was so busy with his veterinary chores.

"Mike's jealous because I can get Eddie to eat more than he can," Eileen told me. "He has a different method. He lets Eddie have some formula first, then some semisolid food. I vary it more. I go with the formula, then some water, then more formula, then the semisolids. For the semisolids, we give him baby cereal, applesauce, and yogurt. We had to call Australia to figure out what baby kangaroos eat. They said the yogurt was important for his intestinal flora. Some people at the San Diego Zoo suggested the cereal and the applesauce.

"But Eddie's really doing well. He hasn't started to hop yet, but he's walking. His coordination is coming along. He chases my cat. Kinda slowly. Roxanne likes him better than the binturong because Eddie pretty much stays in his pouch. I think Ralph's always wanting to ride on Roxanne's back was getting to her."

Eileen swiveled around on her stool to look at Eddie's bag.

"The first night I had him home, I stayed up to watch the eleven o'clock news, then I fed him and I lay back and fell asleep on the couch. He was in his bag on the floor. The next thing I know, I'm hearing this bizarre noise. I couldn't figure out what it was. It sounded like a hacking cough. I jumped up and found Eddie on the floor looking up at me. He had climbed out of his pouch and was screaming for attention. So I fed him and he went back to sleep. Now, every time I hear that noise, I feed him. I'm getting trained."

2

A few days later, down in African Plains, Dan Maloney was inaugurated into the zoo when Puzzles, one of the female giraffes under his care, dropped her baby. Giraffe babies are delivered headfirst from a standing position, so this was indeed quite a plunge—about five feet. But the infant survived the landing. In fact, she was up and walking in an hour. "It was like trying to balance a block on a long pole," Dan said, "but she made it." She was a spindly six-footer, all neck and legs, who was always leaning against her mother, as though in need of the most fundamental kind of support.

The baby was named Michelle, in honor of Dan's wife. "That would never happen at the Bronx," he said. "There, all the animals are named after the major donors." Keepers name most of the animals in Philadelphia—often after other keepers or family members. Dan hadn't planned to honor his wife this way. When Debbie Derrickson from PR arrived and asked what the newborn's name was, he'd started shouting out suggestions, and soon came up with Micky. Another keeper said, "Hey, what about Michelle?" So Michelle it was. "It was kind of a wedding present," Dan said. Mrs. Maloney herself, the graphic artist, was flattered at first. Then photos of the baby giraffe went out across the country on a wire service, with her name attached. That was a little strange, she thought. It was as if her husband had turned her into a giraffe.

——— 3 ———

Dave Wood is senior keeper Bill Maloney's assistant. Thirty years old, he has a scraggly blondish beard, light brown hair, which he wears nearly to his shoulders, an angular face, and a slender body.

Dave's office is across from Eileen's lab in Penrose. On the rare occasions when he wasn't out working with the animals in the garden, Dave would be leaning back in his creaky, roller chair with his boots up on the wooden desk. Keepers would swing by occasionally to hear their day's assignments. He'd have some coffee going, and a pack of cigarettes handy. He'd be in the mood for talking.

"All my life I've known that this was what I wanted to do," he said. "There was nothing else. I don't know what the hell I'd do if it wasn't for this job." He'd grown up in a tough part of Philadelphia called Fishtown. His father was a plumber for the Smith Kline and French pharmaceutical laboratories; his brother is a plumber, too. But Dave always knew he had a different calling, for he had always had a way with animals. "When I was growing up," he said, "if there were thirty kids sitting around, I'd be the one the turtle would come up to out of the bushes."

He first came to the zoo as part of a high-school training

157

program. He was assigned to the Children's Zoo. He went into the Children's Zoo barn that first day, but couldn't find anybody. He yelled out, "Hello? Hello?" but got no response until finally he heard a scratchy voice at the top of the stairs reply, "Yo, Ann!"

"Hello!" Dave responded. "It's Dave Wood. I'm here to start that job . . . "

"Yo! Ann!" returned the voice.

"No, not Ann. It's Dave, Dave Wood."

Finally, Dave located the voice at the top of the stairs. It was a yellow-headed Amazon parrot.

"I thought to myself—here I am on my first interview talking to a stupid parrot. I've got to be crazy."

"That's why we hired you," said Mike Barrie, who had been walking by and caught the end of the story.

"Get out of here," said Dave.

At sixty-four, Bill Maloney was old enough to be Dave's father, and they both might have liked it if he were. If the two weren't out in the garden grappling with an animal, they were in the animal services office, swapping stories and kidding each other. Dave admired Bill more than anyone. "I can react to where an animal is going," he once told me, "but Bill knows in advance." Bill was of the old school, though, and a little shy with me. Dave was just the opposite. He had trouble remembering my name at first (the only things that are sure to stick in his mind, he once confided, are related to animals, particularly elephants). But before long, he'd yell out my name with enthusiasm, even outdoing Dan Maloney, who had called me Sedgman, by labeling me "Sedgmaster," giving me a big handshake or a slap on the back, and making me feel very welcome.

Now, in late October, he was a little more harried than usual. For Xavira was due to deliver her long-awaited offspring any minute, and Bill Maloney had had to go into the hospital for an artery bypass operation, leaving Dave in charge. Maloney had come out of the surgery fine, Dave told me. He had recently been to visit Bill, and found him shuffling up the hall with his IV bottle. Bill joked that he had gotten up early to go the Melrose Diner, his usual South Philadelphia eatery, but the place was closed. So he'd had to walk all the way back.

Dave Wood was used to pressure. Before this job, he had been an elephant keeper, and there is no more dangerous job in the zoo. The Philadelphia has four elephants—two African and two Asiatic—all females. The Africans, Petal and Kutenga, are bigger and fiercer than the Asiatics, Peggy and Dulary. The Africans are also distinguished by their larger ears and more erect posture; the Asiatics have pudgier bodies and slightly humped backs. Like tigers, anacondas, and polar bears, all elephants are proven killers; but unlike the keepers of those other animals, the elephant keepers regularly go into their enclosure with them, in part because of the animals' very unmanageability. The keepers could never persuade them to move from cage to cage at cleaning times without an occasional slap of the hand or a poke from an elephant hook. Also, the keepers have to be able to get close to the elephants periodically to clip their toenails. However, once the bond with the keeper is formed, it is remarkably strong. At the zoo in Hanover, Germany, the elephants refused to come inside from the yard after their regular keeper was hospitalized because of a zoo accident; they wouldn't obey anyone else. An exasperated zoo management had to drive the keeper back from the hospital and carry him out to the elephant house on a stretcher. There he ordered the elephants inside, which they meekly did as soon as they heard their master's voice.

Elephants are notoriously unmanageable, but the zoo's group particularly so. Three of them were raised at the Children's Zoo, and it is likely that the indulgence they received there accounts for the great difficulty the zoo has experienced. But the fourth, the Ceylonese Peggy, who had had the TB scare in September, is no pushover, since she was wild-caught at the relatively advanced age of six. "So she had six years under her belt at being a wild elephant," said Wood. "She's mellowed a bit in the last few years, but when she was younger she was hell on wheels." Even professional elephant trainers have thrown up their hands at the prospect of handling the foursome. Wood brought one of them down to the elephant yard to see what he could do. "That guy left very frustrated," he said. "He couldn't even get within range of 'em." As a final insult, the matriarch, Petal, picked up a branch and hit him with it as he was leaving.

Besides Wood, there were only two other men who could work with the four elephants, the formidable McNellis brothers. The older one, Jimmy, was in fact *so* formidable that around the zoo he was respectfully nicknamed "God." The McNellises had been keepers since the forties, when the elephants were babies. Having grown up with the McNellises, the elephants looked on them practically as their parents.

Dave Wood was a newcomer, and, as such, he had had to work to win the elephants over. Many others had tried, he told me, and they all had failed. Because the job is so demanding, the zoo routinely runs nearly all its keepers through the elephant yard to try to find someone who can handle the pressure. "It's a real test," said a keeper who had tried and failed, "like walking point in Vietnam." Another would-be elephant keeper, a woman named Pat Emery, was nearly killed by Petal. Emery tried to chain Petal up for the night one evening, and Petal suddenly got cranky. She fanned her ears straight out, rolled her trunk into a tight coil, and raised her ivory tusks as though she were drawing a bead on her keeper. That was a warning to keep back, but it went unheeded. With a roar, Petal thrust a tusk down into the base of Emery's throat, piercing a bronchial tube, puncturing a lung, and missing her heart by inches. By luck, the blow knocked Emery through the bars and out of the exhibit. Petal was trying to drag the keeper back in with her trunk and finish her off when the McNellises rescued her. Emery was in the hospital for three months, and had nightmares about the attack for years afterward. She finally left the zoo; she is a housewife in California now.

Chuck Ripka had been trying to work with the elephants for eight years now without success. He still couldn't go in with them without one of the McNellises, or Dave Wood himself if he was around, standing guard over him. As it was, he had had a series of mishaps that might have told him, if he'd cared to listen, that he wasn't making progress. The elephants have cracked his ribs once and his skull twice. But Chuck still wouldn't give up.

Even Bill Maloney, who has an unerring touch with the rest of the animals, had little luck with the elephants. He had thought that things were okay between them until one day a few years ago he was driving slowly by the exhibit in his pickup truck, and the two Africans started pelting him with rocks. Bill got the message. Until then, he had always cut through the elephant yard, even if they were there, to deliver the paychecks to the keepers. Now he always walked the long way around.

Lately Gene Pfeffer had been taking a crack at the job. The hefty Vietnam veteran had been working with the giraffes down in African Plains until Dan Maloney arrived. Gene had been very determined with the elephants, Dave said, but it was too soon to tell if he was going to succeed.

"What's the key to it?" I asked.

"You have to get the elephants' respect."

"How do you do that?" I asked.

"I'm not so sure," he said. "If I knew, I'd bottle it and sell it. There's a point where you have to hold your ground and be prepared to go full swing. You've got to be ready for that. I think attitude has a lot to do with it. You really gotta like working around elephants, and they somehow pick that up. I was attacked a few times in the beginning stages, but the odd thing was, the attacks weren't as severe as they were on other people. It was more like a warning, a test I suppose, than a real attack. They wanted to see what I was made of."

"What happened?"

"The first time I almost got it was one day when I brought them inside, chained them up, and I had gone back into the cage. Petal, one of the big Africans, was facing away from me. She turned around to look at me, and she kept coming and coming until she smacked me with her trunk and laid me against the bar in the cage. I slid down and she just sort of loomed over me. It was scary, 'cause she'd used a tusk on other people. They like to impale you. I was flat on my back underneath her. She coulda got me, but she didn't follow through. She just hung there, looking at me. And that gave me time to slip away between the bars. My feeling is that Petal sorta liked me, but she wanted to let me know that she wasn't going to let me be dominant over her easily.

"You can't be too aggressive with the animal, but you can't be cowardly either. There's this line you've got to walk. I've seen people who are too fast with the hook, and not fast enough with the thinking. Well, that may get them by for a while, but the animal waits for his opportunity. The animal has got to like you, but I'll be damned if I know how to make that happen. If you are too tolerant, you end up with a large, spoiled, dangerous child. If you are too aggressive, you end up with a large animal who is waiting to get revenge.

"See, elephants respect dominance, because dominance is all they know. They are either striving to be the dominant animal in the group or they are following the dominant animal in the group. We all like leaders or heroes. The elephants have their dominance hierarchy, with Petal at the top. Kutenga sometimes thinks she's number one, but really it's Petal calling the shots. She is the alpha female. The keeper going into the group has to become super-alpha. You've got to have total command, total respect. Of all of them. If you have the respect of just one, the next one will come at ya. You gotta have them all."

"If not?" I asked.

Dave smiled again. "They'll kill you."

"How do they do it?"

Dave took a drag on his cigarette.

"The Asians are real good at kicking," he said with a smile. "Real good. They can kick sideways, all the way around, like a cow. Behind, out to the side, forward. I'm talking about the females. The males fight more like the Africans. They use their tusks. They like to knock you down with a kick or a slap of the trunk, and then come down with their heads and squash you into the ground. Or into the wall. They'll lie down on you, they'll sit on you, they'll crush you to the ground with their head, they'll drop on you with their feet. An animal weighing five, six, seven, eight tons—it doesn't take much effort to do something serious to a two-hundred-pound person. A lot of times there have been elephants that have had it in for someone and used another elephant. That person will pass in between and"—Dave smacked his hands together—"the old squeeze play.

"You know, the funny thing is that people think that elephants are these real sweet-tempered animals. In Europe a young bull elephant grew up at the zoo, and everybody got to know it. It wasn't well trained, it was spoiled. So finally they had to tether it indoors. Some girl thought that it was terrible that the elephant should be kept in the building all locked up like that, so she snuck into the zoo at night and climbed in through the elephant house window and went in with the bull elephant to console him because he was so depressed at not being led around. The elephant killed her and ate her. They found her hand and part of her pocketbook.

"But I have to tell you, I owe my life to an elephant. When I was learning to work around the elephants, I was so frustrated that there was a point where I wanted to put a gun to them. Petal was the worst. She was the hardest for me to get around. But now, if anyone wanted to take a gun to her, they would have to get through me, because she is my favorite. And she saved my ass. So I owe her."

He told me the story.

"I was starting to get along with Petal okay, but Kutenga was really busting me. She is the other big African, and she is totally nuts. I keep telling the keepers that I am going to add Valium as part of her regular diet. She is a really high-strung elephant, and she is bad, real bad. When I work around her, and when the keepers do, it is like working on the razor's edge. This is an animal whose eyes are three times bigger than normal. Her temporal glands up on either side of her face are constantly secreting. You can see the hormonal juices coming down the sides of her head. The only thing that keeps her in line is the fear that if she breaks, she is going to get her ass whupped.

"One day we're out in the yard cleaning up and the wheelbarrow got full, and I told Jimmy McNellis, the senior keeper, to go ahead and dump it, I'll wait out here, because we had more to pick up. So I was just standing there by myself holding a shovel when Kutenga came up and shook her head very aggressively. The ears went out, straight out from the head, and she brought herself up taller and the trunk got all wound up tight and she raised her tusks up eye level. And she came up with a jabbing motion at me. All I had was a shovel. So I start whopping her over the head with the shovel. That made her hesitate, but it didn't stop her. She thought about it for a second. I was in a position where there was no place to go. If I had turned to get out of there, she would have been on me. Then Petal, who was at the opposite end of the yard, let out a roar like I've never heard and came flying up at a full charge, and I thought, 'Oh my God, this is it. They're *both* going to get me.' Kutenga looked a little excited about it. Like, 'Oh boy, I got help.'"

Kutenga was mistaken.

"Petal came roaring up and laid into Kutenga like you wouldn't believe. She knocked her off her feet, and when Kutenga got up, Petal gave her two, three fast jabs with the tusks and Kutenga got out of there. And Petal stood between me and Kutenga sideways, purring. It's a low, rumbling noise elephants make when they are content or happy. It sounds like thunder way off in the distance. She stood there until Jimmy came back. Jimmy looked a little surprised.

"I had no idea that Petal liked me that much. Really. This was nine months after she whopped me with her trunk.

"After that, Kutenga and I got along a little better. And then she tested me again. This time, I just—something snapped inside me and I went after her with a pole. It was a telescopic pole that they use to paint ceilings. When it's collapsed, it's pretty solid. I went after her with it. It was like hitting a dog with a newspaper really, like when you smack a puppy for making a mess on the rug. She fought back, but I had the advantage this time, because the pole was longer than her tusks and trunk, so I could hit her, and she couldn't hit me. Eventually, she backed down, and I just went after her and chased her around the yard and tried to kill her with this aluminum pole, which is laughable when you think about it. But it shook her up so much, she never tried it again.

"Since then, I started carrying an elephant hook called an ankus. It's something the Asians use."

Dave got the ankus from his locker to show me. With a thick handle and a sweeping hook, it was an impressive piece of equipment that looked in fact a bit like the top end of a bishop's crook. But it meant business.

"After that incident, the hook did it. I could just grab this thing and Kutenga would back down. It's bad practice, but sometimes I don't even carry the hook with me. Voice command and a smack of the hand will usually do it."

He'd won the elephants' respect.

* * *

Dave had something else in his locker that he wanted to show me. It was a bottle of Great Western champagne. He'd picked it up two years ago as he dashed to the zoo when he'd heard that the Indian rhino, Xavira, had gone into labor. Since the rhinos live next to the elephants in the Pachyderm House, all the keepers there had taken a special interest in the progeny of Billy and Xavira. But none more than Dave, who recognized the historic implications of the birth. If successful, this would be the ninth Indian rhinoceros born in American zoos, and the first in the Philadelphia.

Later, Dave showed me a videotape of Xavira's first labor. Rhino babies are, as humans put it, born with a caul. They are delivered still encased in the amniotic sac. As soon as the baby slides out of the birth canal, the mother whips around, snapping the amniotic balloon. On the tape I could see the baby's four ghostly white little hooves, pressed up against the amniotic wall before it broke.

But the baby's heart had never started beating on its own. Keepers quickly snatched the baby, a female, away from her mother. The videotape showed arms reaching in from outside the frame to drag the baby away. Remarkably, Xavira did not raise a fuss when her baby was taken from her. Possibly she was too tired from her labor. But Dave attributed it to a "mellow" temperament. Keith Hinshaw, already exhausted from staying up all night waiting for the birth, administered CPR by pumping air into the baby rhino's lungs with a crude bellows, and thumping on her chest with his hands. On the tape, I could see Keith's skinny body jerking up and down, over and over, as he tried to revive the rhino. The baby survived eighteen hours this way, but finally stopped breathing and could not be resuscitated. The autopsy showed that she had suffered complications stemming from a uterine infection in her mother. "I don't think the baby ever realized that she'd been born," Wilbur Amand told me later. Xavira herself seemed saddened by the death of her offspring. For weeks afterward, keepers heard what sounded like crying, a nasal whining sound that would go on late into the night and echo eerily around the Elephant house.

That was February 1984. The zoo brought Xavira and Billy together again the following summer. It was the result of that union that the whole zoo was focused on this October. Chuck Ripka, the would-be elephant keeper, had set up an office pool to bet on the date of $2 apiece. More arduous, an around-the-clock rhino watch had been going since October 8, which was sapping the strength and willpower of the organization. The keepers kept an eye on Xavira during the day. They collected urine samples to check her hormone levels, and even milked her from time to time to see if any colostrum was coming in. A video camera had been set up once more in the corner of her stall to monitor her indoors at night; the TV monitor was installed in the conference room upstairs in Penrose. For the staff, that meant a rather grueling process of checking the monitor regularly through the night for any unusual movements from Xavira.

For weeks now, Xavira had slept like a baby—it was the humans who were twitching and moaning. They had installed a cot up in the conference room, but no one got much sleep on it. Who wanted to be the one who'd slept through the historic rhino birth? A manpower shortage made the job even more taxing than usual. Along with Bill Maloney, the curator of mammals, Dietrich Schaaf, was also gone. He left to take the job of general curator at the much-maligned Atlanta Zoo (where, among other embarrassments, the gorillas had been reduced to watching broken television sets). The search was now on for a new curator of mammals, and candidates were trooping by at a rate of one or two a day for interviews with Wilbur Amand. Seeing the procession, the staff had suggested that each of them be required to spend a night on rhino watch.

Still mindful of Xavira's tragic first birth, Dave Wood was determined that this new baby survive. He had slept with the first baby the one night she lived. "We worked so hard for that birth," he said. "Her death was hard to accept. We spend years working with Xavira, checking heat cycles, trying to figure out when to get her and Billy together. Finally, we get them to mate, then to conceive, and then we wait another year and a half for the baby. After all that buildup, all those years of anticipation, the baby dies. I cried when it happened. I get choked up even now when I'm thinking about it. It was really painful. I'm not married. I don't have kids. That's as close as I have ever come to having a child myself. That was my baby."

He took another look at the champagne bottle. "But now Xavira's pregnant again and she's ready to drop. As soon as the baby hits the grounds, stands, and looks healthy, we're going to pop this bottle. And we're going to have a good time, I tell you that."

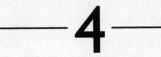

4

Gene Pfeffer won the respect of most *people* right away, even if he was struggling to get the respect of the elephants. At six feet three, 250 pounds, Gene is of elephantine dimensions; he has a five-o'clock shadow that begins at about nine A.M.; and he has a surly look that can cut right through you. When I first met him in August, I tried to ask him a little bit about his family. "Leave them outta this," he said, sternly. When I tried to inquire about his career with the navy in Vietnam, he cut me off. "What is this? My goddam life story?" But he could be gentle, even suave in his way. He liked to flirt with Debbie Derrickson in PR, and when he heard she was thinking of leaving, he was genuinely downcast. "You *can't* leave!" he exclaimed. "You got a *home* here!"

The first thing that you notice about the elephant house is the bars. They are thick shafts of steel, spaced about two feet apart, enclosing the two areas where the elephants pass the night. The bars are no longer straight. They are twisted and crooked as though they have been used to constrain an errant, possibly drunken, operator of the heaviest snow-plow equipment in North America. Gene saw my look of amazement. "Somethin's gotta give," he said with a shrug, "and the elephant's ain't gonna."

At that point, in August, Pfeffer had been working with the elephants for six weeks. Like Chuck Ripka, he always labored under the watchful gaze of one of the McNellis brothers or Dave Wood, if he was around. He spent most of his time feeding the elephants hay, or cleaning up after them with the wheelbarrow. He found the elephants, in this respect, greatly preferable to the giraffes he had previously been taking care of. "Giraffes do it while they walk," he said, "so it goes all over the place." Elephants are much neater, depositing their excreta in tidy lumps called boluses. Gene tended to rank the animals by their bathroom habits. Billy the rhinoceros came out quite high on this scale, since he did his business (as all rhinos tend to) in the same spot at the same time of day. Xavira sometimes enjoyed going in her bath, which was a nuisance. The hippos down at the end of the building were the worst. "They smear their crap all over the walls," said Pfeffer. "That's a pain."

Elephants may be neat, but they are also copious. They eat a hundred pounds of hay a day. Remarkably, the hay is little altered by the experience of traveling through the elephant's insides. The elephant boluses that Gene shoveled into the wheelbarrow often contained a good bit of digested hay that was in every way indistinguishable from the fresh hay that he was laying down for them to eat. (For his 1936 monograph on elephant physiology, Dr. Francis G. Benedict once conducted a peculiar experiment with some zoo elephants to find out precisely how poorly they chew. He sliced up an automobile inner tube, baked the strips in some loaves of bread, and fed the loaves to the elephants. The rubber passed right through; when Benedict fished it out of the elephant's stools, he found no teethmarks on any of the strips.)

Nevertheless, elephants can be very particular about how they are fed. Pfeffer had heard about one elephant who used to apportion her hay into fifty small stacks all around her cage. Then she'd go around and eat them in the exact order that she had laid them down. If a keeper touched one, she'd have a fit. George Myers, down at African Plains, had a similar story about Burma, an old Asiatic elephant, from his days at the elephant house. Burma was very fussy about the way her meals were presented. She used to stand over her keepers with her trunk coiled as they laid out her dinner. The carrots had to go here, the feed there, the greens in a third place. "If you didn't get it right," said George, "she was ready to bang you with her trunk and send you flying out of the enclosure. The trunk was like a cocked fist. One wrong move and *pow*." Keepers sailed out of the exhibit regularly, George told me.

The elephants were out in the yard as Pfeffer showed me around, but the two Africans, Petal and Kutenga, strolled up to the big steel girder that barred the entranceway to the elephant house and stared at us. I got the feeling they wanted to know who I was. They were completely unabashed, and stood like teenagers on a street corner chewing gum and watching everybody go by. Their long trunks dangled down to the ground like garden hoses.

Peering out from the elephant house, I took a good look at them, too. Dave Wood occasionally brings blind children in to touch the elephants with their hands. (The elephants would never try anything when Dave is there.) "That's the only way they can sense an elephant's size," he says. Until now, as I looked at them close up on the other side of the girder, I felt I, too, had been blind to their hugeness. Their heads alone were massive, like great boulders; their bodies were so vast that it nearly required the entire expanse of the twelve-foot-square passageway to let them inside, and their ears were so big that the *Dumbo* idea of using them as wings did not seem so unreasonable. (Curiously, the elephants' feet, which bear the eight-ton load, are structured so that the heel rides on a thick wedge of fatty tissue; in a sense, elephants walk on tiptoe.) I was surprised to see from this distance that their (proportionately) little eyes had six-inch-long eyelashes, as though Walt Disney had gone to work to clean up their image.

"You gotta watch Kutenga," Gene told me. "When she gets wired, she leaks that stuff down the side of her head." Kutenga was oozing hormones now. "Petal, she's all right, but she can be sneaky. With Kutenga, if she's gonna getcha, she's gonna go getcha. Petal, she sucks ya in, then she gives ya one of them squeeze numbers." The two of them stood there. They had started to swing their trunks back and forth in perfect coordination, like a pair of synchronized grandfather clocks. "Stand back," Gene said. "They might start throwing things."

I snorted at that. What could elephants throw? And we twenty feet away? Then the first pebbles started to shower down on us. Petal had started it, and Kutenga had quickly gotten into the act. They rooted around in the dust with their trunks for pebbles and small stones, and then like softball pitchers warming up, they lobbed them toward us with high, arcing tosses.

Gene recommended that we ditch it. "They don't respect me," he grumbled as we retreated. "They'd never do that to Dave or Danny or Jimmy. Petal wants to draw me out so she can get me. They don't have any respect for me at all. I know it, and they know it, and they know that I know it."

In early September, Gene was acting positively sprightly. "Got me a whip," he said. "A buggy whip." He showed it to me, a long strip of leather on a short wooden pole. It looked nasty to me, but I suspected it didn't seem quite so frightening to a pachyderm. "I smacked her with it, too," he said.

"Who?"

"Petal. She was throwin' sand at me, and I hollered at her. She kept on, so I hit her with it. She backed right off." He seemed pleased at that. "It got me some respect."

"Was it your idea?" I asked.

"No, I was talkin' to Jimmy McNellis and he said ya gotta get somethin'. He said how 'bout a whip? I've got a couple of 'em. This and a metal switch."

"Does Jimmy use one?"

"Jimmy? Are you kiddin' me? Jimmy don't need no whip. Jimmy don't need nothin'. He's got their respect. When you got respect, you don't need a whip."

"Is the whip helping?"

"Oh yeah. Petal don't challenge me so much. After I hit her with the whip, she's gone along with everything I said. It used to be she'd keep bumpin' me with her trunk. Now, she comes over and sniffs me just a little bit, but she don't hardly touch me."

He had just come back from ten days vacation in the mountains. Curious about elephants' fabled memories, I asked Gene if they remembered him.

"Oh sure." He said he could tell because they didn't pay him much attention. When a complete stranger comes into the yard, the elephants can't stop staring, as I had found out last month. "They're checkin' him out, what's this guy about, what's he goin' to do?" Gene said. "They're always watchin' you."

Since Gene had gotten his whip, the only trouble had come with the introduction of some heavy machinery across the walkway in the new World of Primates construction. The elephants had finally seen something bigger than themselves— the huge tractors for molding the earth—and they had freaked. "Kutenga got all upset," he said. "Peggy trumpeted. Kutenga took off to get that son of a bitch. When somethin' like that happens, they look to Jimmy and Danny for protection. They respect them that way. They're the paternal ones, they're the bulls.

"The elephants know what's happenin' all the time. They know when I'm feelin' bad, and they try to take advantage of it. When Jimmy and Danny are close by, they behave themselves, but if they're not, they act up. Like another time when Petal was chuckin' sand at me. I wasn't feelin' good. I got the metal switch and laid it on her tusk. I said, 'You keep throwin' sand, and you're goin' to wear this thing.' Well, Petal went over to Kutenga and I could see 'em like whisperin' to each other. I know she told Kutenga about it, because Kutenga backed right off. She went, 'Whoa!'" Gene threw his arms up with a startled expression.

He had to get back into the yard, and I hung around the fence to watch him for a moment. Kutenga and Petal stalked about, their trunks swishing back and forth. Peggy and Dulary strolled together along the barn wall. Huge as Gene was, among the elephants he looked like a child. As he had said, the elephants paid no attention to him. They roamed to and fro around the exhibit, completely oblivious to him and his little wheelbarrow as he went about scooping up the boluses.

Because there is literally no spring in their step, elephants are the only hoofed animals that can never leave the ground. Still, far from galumphing stereotypes, their massive shapes can attain a beautiful, nearly balletic grace, and, when they want to, they fairly fly across their exhibit. Gene, by contrast, seemed earthbound and clunky. The elephants occupied a separate dimension, another plane of reality. Gene worked the same space as the elephants, but in another way he didn't intersect with them at all. The elephants loomed over him like gods. He sought their favor and feared their wrath.

5

Fall was coming on hard now. The air was turning chilly, and the shadows ran long across the yards. In the outdoor gorilla yard, Chaka and the five-year-old Jessica had a delightful time chasing the maple leaves that blew around. The gibbons, stuck away under the monorail stop, nibbled on them like potato chips. And the cheetahs liked to flop onto the beech leaves that gathered in the corner of their exhibit and watch them flutter into the air. But the giraffes, always fidgety, seemed unnerved by the locust leaves that skittered across the yard; they sometimes started at the sound, their ears twitching anxiously, as if they had just seen a mouse.

Some of the gusts of wind had an arctic snap to them, which must have pleased the polar bears after such a long and torpid summer. Occasionally, they sat up on their rock and sniffed the air, as if they were trying to smell the coming snow. The breeze ruffled the hairs of the hoofstock, which pranced about with new verve to keep the blood flowing. And it tousled the new shaggy coiffures of the wolves and the camels, who were busy growing their winter coats.

As the month went by, and the leaves curled on their branches and spiraled to the ground to lie in heaps, the veil of illusion that horticulturist Chuck Rogers so conscientiously wove

175

about the zoo fell away and you could see that this garden was, at heart, a fortress. The chain-link fences, the dry moats, the cement walls were all laid bare.

From the slight rise by the Small Mammal House, the highest point in the zoo, you could see clear across Bird Lake to the far picnic grove. The workmen had started to set out little orange flags there tc mark the spots for the new construction that had been sponsored, as it were, by K'Bluey. Viewed from this vantage point, the zoo exhibits all ran together. Without the walls of vegetation to screen them off, you could see how the African Plains pushed up against the hoofstock alley, and how Bear Country pressed hard on Bird Valley. It was like the backlot of a movie studio.

In another sense, though, this was a grand reunion. For many of these animals, the Australian marsupials, the African antelope, the South American bears, had not walked the same plot of earth since the tectonic plates broke up in the Triassic period. In the zoo they were back together once more.

With the leaves down, you could also see that some of the zoo interlopers, chiefly the squirrels, were overrunning the place. The squirrel nests were everywhere, big pendant canisters clinging to the branches. The trees were heavy with them—four, five, six in a single tree; they drove Chuck Rogers to distraction. "Rodents love a free lunch," he grumbled. "But they're terrible for the trees. You can only take so many; after that they start to do too much damage. They chew up the bark and girdle the trees. Or they build nests in hollows, which isn't too good either. They get into the greenhouse and chew everything up, and they sneak into the flowerbeds outside, too. We have some pest control, though." Chuck smiled at the thought. "Frank Russo comes out with his twenty-two, but I'm not supposed to tell you that."

As the temperature went down, many of the animals went inside. Most of the birds and reptiles were pulled; the mammals could decide for themselves when it was time to go inside, since many of them had immediate access to indoor quarters.

At the end of the month, a crew yanked a couple of crocodilians from their yard behind the Reptile House, the one that Prickles had strayed into over the summer. Wrestling with alligators is ancient sport, but these folks weren't taking any chances. Vigorous and outspoken as a five-hundred-pound American alligator can be with its snappish jaws and thrashy tail, it can be almost completely immobilized by some electrical tape. A little six-foot caiman had already received the treatment when I arrived. With his snout taped shut, he looked not just powerless, but really stupid. It was as if he had been in class, and the teacher had said, "One more dumb remark out of you, buster, and—I'M GOING TO TAPE YOUR MOUTH SHUT!" So there he was now, stretched out inside the fence, his snout encircled by silver tape, waiting to be shipped out as peaceably as a rolled-up rug. A big American alligator was next. Denise Robinson, Dave Wood, and Bob Pittman stood around thinking things over. Curator of reptiles John Groves was urging them on.

Groves is a lanky fellow with a small moustache. He is the son of the curator of reptiles for the Baltimore Zoo; as a herpetologist, he himself was entirely self-taught—he had no academic credentials.

He had given Bob Pittman, a rather slender substitute keeper, the task of hauling the twelve-foot alligator from his pool. Bob did not relish the assignment, and he timidly plopped the rope into the water well away from his quarry.

"Bob!" everybody yelled.

Finally, on about the tenth attempt, Bob succeeded in roping the 'gator. Dave helped slide the lasso down to the animal's middle, and then the two of them hauled on the rope to drag the beast to shore. It was like lassoing a log, but somehow they managed to bring him in. Then came the tricky business of the tape.

The three keepers looked at each other; no volunteers.

"Oh, I'll do it," said Dave finally. There was a look of relief in the eyes of Bob and Denise.

"Okay," yelled Groves as Dave moved into position. "Drop the cloth over his eyes, then pounce on him from behind!" Dave stood over the animal and placed the cloth, about the size of a large napkin, over the alligator's eyes, then quickly drew back his hands. Surprisingly, the alligator made no move to shake the cloth off. Instead, he went perfectly still, almost limp. If the lights are out, he must have thought, it's bedtime.

Dave swung a leg over the alligator and lowered himself down as though he was settling onto a precious antique. "Sit right down on him," yelled Groves, "he won't break." Dave dropped himself on him like a professional wrestler. The 'gator remained still. Dave crouched forward and slowly extended his hand toward the snout like a strangler approaching his victim in the dark. Then in a flash, he pounced on the 'gator's jaws and squeezed for dear life. This did try the alligator's patience, but there was nothing he could do about it except hiss. Which he was doing fiercely. "Hurry up with that tape, would ya?" said Dave. Denise set a paper towel on the alligator's face so that tape wouldn't stick to its skin, then wound the tape around and around the 'gator's snout. His teeth protruded menacingly around the edges.

"Yeee," she yelled. "I'm going to lose my finger!"

"Just do it, Denise."

The job completed, she patted the tape down to make sure it would hold, and took a moment to inspect the alligator's incisors. "Oooh, those teeth are sharp!"

"Damn right," said Dave.

He eased off the creature a bit, and the alligator started to writhe. "Watch out! He'll jump!" shouted Dave. Everyone leaped back. Dave squashed himself back down. The cloth slipped down the beast's snout, revealing a pair of angry eyes. Denise hastily pushed the cloth back up.

Straining mightily, Dave, Bob, and Denise tried to hoist the 'gator over the chest-high fence and into the back of Dave's Dodge Ram truck to drive him around to the indoor exhibit. There weren't many good handholds, and alligators are slimy. After much effort, they managed to get him on the top of the fence. Just as he was poised up there, he suddenly started to slide backward. Some visitors who had gathered around for a better look—as people at the zoo always do whenever the keepers are doing anything with the animals—all scattered. The 'gator hissed furiously. But Dave managed to steady him and tip him back up.

"What a *butterball!*" Dave Wood gasped as the three keepers struggled to heave the animal into the truck. Finally, John

Groves put his back into it as well, and the alligator toppled into the pickup, where it thrashed its tail angrily and let out hisses from between its clenched, taped jaws. After the fatso, the slender caiman was as easy to move as a handbag. Dave popped him in practically by himself, and then drove off to stash them both in the Reptile House for the winter.

When I saw Groves later, I complimented him on his efforts with such a pair of ferocious reptiles. "Nah," he said, "they're just a couple of pussycats."

The camel concessionaires Tim and Tracy Hendrickson had packed up October 1 to go back home to Huntington, Indiana, for the winter.

Independent contractors, they owned four camels, and they lived with them in a little trailer out in back of the TreeHouse. They were the only human residents of the zoo. "It feels like a prison, but it's a prison where we've got parole," Tim said. A slim little man, he wore a baseball cap that read CAMEL across the front and obscured much of his face, leaving little but his Hoosier accent. "There are lots of noises at night, the cats roarin' and the sea lions barkin', but we don't notice 'em anymore. We used to hear the trains runnin' all the time, but now we don't even notice that. A friend came the other day and a peacock was screamin'. He said, 'What the hell is *that*?' It's a helluva strange noise if you never heard it before, you know. But I'd screened it out. I didn't know what he was talkin' about."

It had been a long summer for the Hendricksons. They spent all day every day loading zoo visitors onto the camels' candy-cane–colored seats, then leading them slowly around a circular dirt paddock, and hauling them back off again. While Tim was leading one camel, he was trying to keep the others from rolling in their dung, which, for the dromedary Adelaide at least, was a favorite diversion. It was hot work. "The camels go thirty minutes without water," he said, "but I can only go ten without a Pepsi."

Tim had been a welder by profession until one day he got the bright idea of using a mountain lion in a magic act. He bought one from a dealer and, by means of a secret compartment, made the mountain lion appear and disappear from his cage. "The trick was to keep him quiet," he said. The experience led him to believe that he might be able to work with animals for a living. So he tried camels. Camels, he admitted, can be ornery. They kick in all directions, even up. One of his camels could hit a rafter ten feet in the air. Tim had been hit by the Bactrian (two-humped) Mr. Hooper recently, and he had a bruise on his thigh as big as your fist. He'd been giving him a shot of penicillin twice a day. Mr. Hooper had taken it pretty well, but Tim had been guarding himself with a heavy barrel so the camel couldn't do too much about it. Things were going so smoothly that one morning Tim didn't bother with the barrel, and as he put it, "POW! I'm layin' down moanin'. Tracy came in and said, 'Oh, did he kick you?' I said, 'Yeah, that's why I'm layin' down moanin'.'"

Tim was lucky. Sometimes camels knock you down so they can sit on you. This maneuver isn't quite as devastating as it is when elephants try it, but it is an unpleasant experience nonetheless. Tim was thinking of trying to keep some elephants, but was discouraged by the fact that although camels can get their own ideas after being locked up in the barn at night in Indiana, they just kick at the walls in frustration; elephants would push the whole barn over.

Tim did love his camels, though—particularly Sheeba, a snaggle-toothed dromedary with a particularly dopey expression. The affection was returned. Tracy was a little jealous of Sheeba's affections. "She protects Tim from me," she said. "If we're yelling at each other, she comes right up between us to keep me back. Tim's funny about it. When people ask if he's gonna breed Sheeba, he says, 'Breed her? I sure will, soon as I find me a ladder.'"

The Hendricksons kept their menagerie in the middle of farm country in Indiana. Besides the camels, they stocked a few wallabies, and a monkey or two, but the camels got all the attention. Tim was out in front of the house one day when a car zoomed past the barn, then came to a screeching halt and backed up. The man approached Tim, pointing. "I just have to know one thing. That's a camel, isn't it?"

"Yessir," said Tim. "That's a camel." Thus reassured, the man got back in his car and drove away.

This winter, Tim had plans to put his camels in a Nativity

scene at Christmastime, a lucrative sideline for a camel owner. And he had contracted to transport some animals across country in his trailer. Other than that, he was going to take it easy. "Restin' up for spring," he said. They'd be back at Easter, the first big weekend of the year.

6

Debbie Derrickson never knew when Bill Donaldson was going to loom in her doorway, sing out a "Hi, how are ya," stroll in, sit down, suck on his pipe, and start telling stories, pretty much regardless of what she was doing or who was in her office.

"I must have told you about Bob Snediger," he began one time when I was there. "Curator of reptiles over at the Brookfield Zoo in Chicago." Debbie and I looked blank. "Bob got bitten by snakes so often, he lost all sensation in one hand. He was a chain smoker, and sometimes you could smell the flesh burning."

He shook his head, dismissing poor Snediger from his mind.

"Or Grace Smiley. Oh, they let 'em all out at Brookfield. She thought the snakes talked to her. Years back, she trained king cobras for the movies. Until one of them bit her, and she died."

"Chancy things, snakes," I said.

"Yes, but chimps—now there's a strange animal. I remember a guy named Skelton at Toledo who had a male chimp there that he was really close to. Skelton also had a temper. Skelton'd cuss somebody out, and there the chimp would be—his hair all fluffed up in an anger display, pounding on the glass and stamping his feet. They were a pair, those two."

He thought for a moment or two as he puffed on his pipe.

"Oh, but now Joe McHale, he was a keeper at the Lincoln Park Zoo in Chicago, and he was something else again. He lived with his mother in a house that was littered with TV sets. Must have had a hundred of 'em in there, and all these funny brands like Dumonts and ancient Zeniths. He collected washing machines, too. I went over there one time, and he said, 'Bill, I have something to show you.' So he took me down to his basement to the coal bin, and I don't see anything. It's all black. But then my eyes adjust to the light, and I start to see these eyes, hundreds of them staring out at me. He had a whole flock of penguins down there, all of them black with coal. And he drove around in a Checker cab. His mother was a little daffy, too. She kept a wolf on a leash. I used to drive around with them in Cincinnati. He had some big birds in the back, storks, I guess, and he'd keep a supply of raw fish at his fingertips when he drove. And every once in a while he'd pick a fish out and, with a little flipping gesture, he'd toss it over his shoulder to the storks. He himself used to carry his teeth around in his pocket.

"My friend Eddie Maruska at the Cincinnati Zoo told me once about a keeper named Lester Tidwell who grew up in the mountains of Kentucky, a real backwoodsman. Lester went in the cage with a mandrill. Now, a mandrill's teeth are about this long." He held his fingers two inches apart. "You've probably seen them. I wouldn't go in there, and Ed was horrified. So the mandrill grabs Lester and starts shaking him. 'Well, they don't bite,' says Lester between bounces, 'but they sure shake the crap out of you.'"

Donaldson doesn't usually laugh at his own stories, but he laughed now.

"But one of the funniest stories I ever heard was about the time that the gorilla got out in Brownsville, Texas. The gorilla was stuck out on an island there at the zoo, but some workmen had left a boat out there, and the gorilla climbed in and pushed off and drifted to the far shore, where he jumped out and scared everybody to death and they all just *scattered*. Don Farst is the director of the zoo, and he ran out with a dart gun, but the gorilla just picked out the darts and threw them back. They'll do that, you know. And he went charging off and ran up this hill. Well, there's a lady there with a baby on the far side, and she saw this gorilla coming at her and she took off and left the baby sitting there. A keeper ran up and grabbed the baby and threw him through the window of the souvenir stand. The kid was okay and they finally cornered the gorilla, but it was wild there for a while."

He paused for a second, then he turned businesslike. "So how are things coming with the rhino publicity?"

"Oh, just fine," said Debbie.

They talked about different ways of handling the press's inquiries about the imminent birth. After a few minutes of detailed discussion, Donaldson got up with a groan and strolled out, leaving Debbie wondering if she had, without realizing it, just attended a meeting.

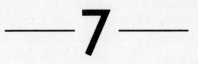

7

A slow loris was spread-eagled on the operating table at the Penrose Annex. "A very slow loris," said Mike Barrie. He was peering in the loris's mouth.

Loris means "clown" in Dutch, and it is normally a rather comical creature with its big bug eyes, button nose, and skinny little brown body halfway between that of a monkey and a cat. Taxonomically it is of the prosimian suborder of primates. The loris is called "slow" because of its stealth as a nocturnal hunter in the jungles of Southeast Asia. In their darkened exhibit behind glass in the Small Mammal House, the lorises moved like cats stalking mice even when there was no prey in sight.

This one, Cloris, was very slow because she was very sick. Some pus oozed out of her mouth, and she emitted ghastly gagging sounds as she breathed. Mike poked gently at her jaw with a tongue depressor. "This jaw doesn't move," he said. "And her tongue is real swollen."

"She's blitzed," said Dave Wood, who was holding her down.

"She looks like you on Saturday morning," Eileen Gallagher said.

"She's got facial paralysis on the left side," Mike continued, "and lingual paralysis. I'd say she's stroked out."

"Can she grip with her left hand?" Keith asked.

"I can't tell," Mike replied. "She's in tough shape."

Keith prepared some ketamine, a tranquilizer, and injected it into Cloris's rear end. The animal grabbed onto the edge of the table with her right hand. Then she relaxed the hand and turned it palm up. The fingers were long and delicate.

Dr. Carl Tinkelman stepped up to have a look in Cloris's mouth. He is the zoo dentist. He also services Main Liners at his office on Fifteenth and Locust streets near Rittenhouse Square, a fancy area of downtown Philadelphia, but comes in to the zoo on a volunteer basis on Tuesday mornings. That was when the vets scheduled their major operations. While the animals were immobilized on the table, Tinkelman took the opportunity to do a checkup, clean off plaque, and pull some teeth if necessary.

Tinkelman is a handsome, well-groomed forty-year-old with perfect teeth. He wore a tattersall shirt and pressed blue jeans that stood out in a roomful of khakis. The zoo had first called on him ten years ago when a tiger needed root-canal work. That operation went well, and the zoo had used him on an emergency basis until two years ago when they put him on staff.

"Animals' teeth are structurally the same as humans'," he once told me. "They're just larger." He had replaced with false teeth the eyeteeth of cats and bears who broke them chewing the bars of their cages; he had inserted a plastic tooth into an otter brought to him by the Pennsylvania Game Commission. But generally the problems occur in the gums, not the teeth. "They don't get cavities," he said, "because they don't eat sweets and carbohydrates. They do get abscesses in the gums because their gums don't always get the stimulation they need. The cats, for instance, just get meat, they don't get anything to gnaw. Give them leather and they're through it in two minutes. But we're working on improving their diet." It was an urgent problem, for the progression was all too familiar: The gums swell, then bleed, an abscess forms, some teeth rot, the animal stops eating, and then serious illness sets in. A timely trip to the dentist can save an animal's life.

For surgery, Tinkelman waits till the vets put the animal under, then he props its jaws open wide with a spring, and performs the work pretty much as though the animal were a human in the chair at Rittenhouse Square. He hasn't had any serious problems with animals coming out of anesthesia a little earlier than expected, although a tiger did take a swipe at him once as he stirred in his slumbers. "I just moved away until they brought him back down again," he said. His main difficulties with his animal practice are that the camels don't open their mouths wide enough, and that cats have bad breath. In general, he is grateful to view nature's handiwork from close up. "The teeth are so beautiful," he says, "it's a joy to put my hands in their mouths."

Just last week, Tinkelman had been filmed pulling some of Cloris's teeth for a TV show on zoo life. "You know, Carl," joked Keith, who had swung around by now to take a look in the loris's mouth, "I think you're going to have to go back on TV and make a correction."

Keith picked at Cloris's mouth with a tongue depressor. The teeth popped off her jaw like corn off a cob. "The whole jaw is rotten. Teeth are falling out right and left here."

"Her pulse is irregular," said Eileen, who had been monitoring it.

"We're going to have to take out a canine tooth," said Keith. He prepared some Cetacaine, a local anesthetic similar to Novocain.

The loris made more gagging sounds.

"Eeeegggh," said Eileen.

"You going to try a tracheotomy?" asked Carl, referring to the possibility of opening up a hole in the loris's throat to breathe through, since her trachea seemed to be clogged.

"It's not worth the heroics," said Mike.

"There is not a lot of room to work in here," added Keith. He poked around a little more. "Jesus, the whole mandible is soft."

"So what's happening?" I asked.

"She's dying," said Keith.

"She's got a bad abscess in her jaw that's probably cellulitis," said Mike. "That's an inflammation of the cells. She's also got bad cataracts, and she's seven years old."

I looked at him uncertainly.

"That's fairly young for a loris."

Wilbur Amand strolled in to see what was going on. "This guy eating?" he asked.

"We thought so," said Keith.

Wilbur spotted the dentist. "Well, thank God you're here," Wilbur said to Tinkelman.

"We don't need a dentist," said another lab assistant, who was looking on. "We need a priest."

"We just got a male up from Duke," said Wilbur. The university had a breeding colony of lorises. "I guess we'll have to get a female, too."

The vets have developed a lot of terms for dying, since they see so much of it at the zoo: "vaporlock," "flameout," "crash-and-burn," "going to the Great Exhibit in the Sky." "I'm not a religious person," says Keith. "I try not to make too much of it. If death breaks you up too much, you ought to think about another job." Keith and Mike do form attachments to zoo animals, but they try to limit them. And so now they pull back from Cloris as she flames out.

Sensing the mood of the room, Wilbur changed the subject. Despite his peppy demeanor this morning, he had taken the graveyard shift on the rhino watch last night. Xavira's behavior was unchanged, he said. He'd checked the monitor at one o'clock after watching Johnny Carson and then again at four. "She hadn't moved a muscle." And Xavira's appetite was still strong. "She's like a vacuum cleaner," said Wilbur. "I had to search both exhibits for a blade of grass." Unable to sleep himself, he had stayed up most of the night writing up reports. "It was great," he said. "But Dolores is not going to be happy." Dolores was his secretary—she'd have to do the typing. Wilbur bounded out.

"Uh-oh," said Keith. "Her leg is stiffening."

That could be a sign of imminent death, but the loris relaxed again. Mike injected some fluids into the animal's neck, and took a blood sample. Then he carried her over to the incubator in the next room. He set the dial for eighty degrees. I reached in to stroke her fur before he closed the door. It was thick and bushy, like sheep's wool. Cloris didn't respond to my touch, but lay there stiff as a corpse with her tongue hanging out.

"I give her a couple of days," Keith said when Mike came back.

"That sounds about right," said Mike.

I brought up a more pleasant subject. "How's Eddie?"

Mike brightened. "Oh, he's up around seven pounds," he said. "He's growing by leaps and bounds." Then he turned his attention to a small bristly pygmy hedgehog tenrec that was curled up in a ball on the table. It looked like a cactus. "I thought he was about to die," said Keith, "then I found him running on his exercise wheel . . ."

Keith Hinshaw is a veterinarian by blood. His father was the chief veterinarian for the Arizona State Racing Commission; his father's twin brother, Keith's uncle, was a horse veterinarian in private practice. "I was never really tempted by horses," Keith says. "I have a tendency to fall off at high speed. A horse once bucked me and I came down on the saddle horn. You know, I used to weigh three hundred pounds." Another time, he was racing his brother on horseback across a field when a deer cut across his path and caused his horse to tumble down a ravine.

The veterinarian's art is a complex one, and there is a ton of information to master. From the University of California at Davis, Keith obtained a Bachelor of Science degree *summa cum laude,* a Doctor of Veterinary Medicine (school co-medalist) degree, and a Master of Preventive Medicine degree. He also spent time at Davis's California Primate Research Center, where, among other things, he repaired the often gruesome damage sustained by chimps in dominance fights. To keep up with new developments in veterinary medicine, he invariably has a three-foot stack of current veterinary journals on his desk: the *Journal of the American Veterinary Association,* the *Compendium of Continuing Education for Veterinary Practitioners,* and eight others. "I think of the animals as extremely sophisticated machinery," he says, "but it's machinery with no warning lights to indicate if anything is wrong. They don't give you any helpful hints. Your job is to keep them up and running, but everything is against you."

There is an air of the master mechanic about him. Everything about him is regular, precise, but he seems a little tired from overwork. He lives with his cat, "your basic domestic shorthair," named Bozon (after the subatomic particle) in a house in Manayunk, twenty minutes from the zoo, in a slowly gentrifying section of Philadelphia. The house is a stucco-sided, boxy place with a slim yard behind it. Although Keith had owned the place for two years, it was still almost completely unfurnished—just bare walls and bare floors. There was a card table to eat on in the dining room, and on one wall hung the only decoration in the house—a poster of a wineglass filled with strawberries. "I had three priorities," he explains. "First to buy the house, then to save two months' salary, and third to start an IRA." He was just getting going on the IRA, so the decorating had to wait.

However methodical, Keith also has a certain delicacy. He had met his current girlfriend, Becky, one Saturday at the zoo. Becky is the sister-in-law of one of the lab technicians upstairs at Penrose. Vibrant, with bouncy blond hair, she had been taken with the slender fellow in khakis who was attending to some marmosets—small, furry primates about the size of kittens—before going off to give a talk to the Feline Club. A social worker herself, Becky could sense the tenderness that Keith put into his work. She was drawn particularly to his hands, the way they both calmed and restrained the helpless little animals. He had just the right touch, she thought: light enough to keep from hurting the marmosets, but firm enough to keep them in his grasp. "He was so caring," she said, "and his work was so hands-on."

Keith had been struck by her, too, but he was too timid to act on his feelings. She had to send him her phone number, although he had it already. The two soon realized they had a lot in common. At the mental hospital where she worked, she felt the same attachment to her patients that he had to his animals. They both administered to the helpless.

And sometimes the veterinarian is the one to feel helpless. Since the animals are so secretive about their illnesses (particular the smaller ones—larger animals like elephants can be more open about their ailments), Keith often feels that he is operating by remote control. He has to rely on a variety of indirect

methods to establish the most elementary facts about his ani-
mal-patient's condition. Even a precise measurement of an animal's weight is often very difficult to obtain. No one knows
how much the elephants weigh, for instance. There is a truck
scale to weigh incoming shipments by the back entrance to the
zoo, but it is often out of order, and it requires the elephants
to be in a truck to be weighed. So Keith has to rely on more
subtle methods, sensing a slight hollowness under the elephants' temples when they are thinning down, or an unusual
concavity to their haunches.

Another factor extends the divide between man and animal. If a vet pries too much into an animal's condition, the
resulting stress alone can damage it more than the suspected
disease. There are many stories of animals getting frightened
and smashing into walls, or, like Snyder's tigers, dying from
the surge of adrenaline that comes with shock. So the vet must
constantly worry: How far should I go? More complicated still,
of the animals that it is safe to touch, some are soothed by
human contact, others ruffled by it. Keith has learned to re-
frain from petting the camels as he tends to them, for that
frightens them more than anything; but wolves must be man-
handled to be controlled. One solution, of course, is to knock
the animals down with anesthesia before working them over,
but that carries its own hazards. One common anesthetic, ketamine, is a version of phencyclidine, better known as "angel
dust": It's not the sort of thing you want animals taking into
their bloodstream on any regular basis. It works by "dissocia-
tive anesthesia"; it doesn't kill the animals' pain so much as it
changes their outlook. The animals still hurt; they simply don't
care. The moments preceding the knockdown can also be highly
stressful. The vets have to use up to twice as much anesthetic
as the animals' body weight might require, since their absorption rates are so greatly increased by the stress of the knockdown itself.

Mike Barrie regularly had to knock down Gira, the female
African lion. Lacking the male's splendid mane, Gira looked
like an overgrown housecat, but she could be murderous. She
had a raw spot at the base of her tail stemming from an operation to cure a uterine infection. The stitches must have itched,

for Gira had been regularly rubbing the base of her tail against the wall, and she had opened up a nasty sore. Painful as it must have been, Gira kept scratching it. The vets had to bandage the wound to let it heal. But that meant changing the bandages weekly, and that, in turn, required Gira to be anesthetized. Mike used a blowgun, a long copper tube that looked like something the pygmies might employ, to discharge the drug-laden darts.

After a few weeks of the treatment, Gira had understandably grown suspicious of anyone approaching her holding a long tube to his lips. She snarled, pacing angrily back and forth in her cage; then, as Mike drew close to fire, she sprang at him and drove her massive front paws hard against the bars. The ancient Carnivora House, lined with tiles, rang with her fury. As she turned sideways, Mike puffed out his cheeks and blew. There was a *ffft* sound, and a tiny dart, trailed by a little orange tassel, flew into her cage and smacked into her shoulder. The blow must have stung, for she swung around at the impact. Then she slowed down a bit. Mike fired two more darts. She slumped to her haunches, and her eyelids began to droop. Finally she rolled onto her side and lay still. After poking her with a stick to see if she was completely unconscious, Mike and Keith climbed into her cage to change her bandage. The tassels still clung to her hide like ornaments from some primitive ritual. When they were done, they injected her with an antidote to the ketamine, and Gira came slowly back to life.

Even when they are able to get close to an animal, the vets often find themselves just as befuddled as if they were looking at it by telescope from a hundred miles away. For all his veterinary degrees, Keith is still dumbfounded quite regularly by some of the strange symptoms an exotic animal might present. Veterinary schools don't teach much exotic medicine; they concentrate instead on the dairy, agricultural, and pet animals that form the core of their graduates' business. Consequently, Keith often navigates around inside some of the stranger animals by thinking of them as cousins of the more familiar barnyard species. To understand a zebra, they think of it as a horse; to deal with a camel, they consider a cow; to address a peacock, they imagine a chicken. Without such analogies, they might be stranded. "There's a helluva lot more that we don't know about exotic species than that we do," says Mike.

In the face of such confusion, Keith relies on a clear knowledge of first principles, and an orderly approach to further investigation. Generally, he proceeds from the outside in. Reports from the keepers that an animal is behaving strangely cause him to take a look. If a knockdown is required, he begins by palpating the anesthetized animal—feeling over its body for any unusual swellings or protuberances. He does other sensory investigations, taking its temperature and weighing it (if possible). Probing deeper, he may take X rays, and samples of blood or serum. When the evidence is in, he looks for abnormalities, assuming that the norms for the species have been established. From the abnormalities, he makes a diagnosis and suggests a cure.

Most cures, however, don't succeed. The aspect of his work that is hardest to bear is the fact that few of his ministrations are of any use. Most of his patients die, for in most cases the vets don't know that the animal is sick until the disease is well advanced and probably beyond curing. By then, too, many cures, like open-heart surgery, are too expensive. Many others are simply unimaginable. So the vets often end up in the position of the parish priest—giving the patient comfort as he faces his inevitable end.

This is hard for most vets to accept, and Keith is no exception. Dave Wood says Keith once labored in surgery for four hours over a cherry-tipped mangabey, a species of monkey. "It was an old surplus male," Dave recalls. "He was sterile. And he had just about everything wrong with him. Others might have euthanized him, but for four hours Keith gave it everything he had." It was no use. The animal died. "Afterwards, I saw Keith beating up the walls with his hands."

When I asked Keith about the mangabey later, he said that he didn't hit the wall all that hard. "That's not good for your hands." He usually threw things. What kind of things? He smiled. "Whatever's around." Humor is his salvation.

Leslie MacGregor, a young veterinary student who spent some time at the zoo, tried to explain Keith's position. "You always want to save the animal," she said. "When you work in a small-animal clinic, you always want to put the dog back in

SO. ST. PAUL PUBLIC LIBRARY
106 3RD AVE. N.
SO. ST. PAUL, MN 55075

the owner's hands and have them go home happy. But with dogs, if you fail, there are always more dogs. For every dog that comes out of surgery, there are five more in the waiting room waiting to go in.

"It's not like that at the zoo. You have a different relationship with the animals. Here, there is no owner. You as the vet are the owner, in a way. You're the caretaker, and there's a higher responsibility involved. You're the guardian of this animal. This animal has come from the wild and been entrusted to your care. It belongs to the world, and it's been delegated to your responsibility by all mankind. The last thing you want is to see an animal die that might have been saved.

"And there is another thing," she went on. This was something that everyone was thinking about with the rhino coming due. "There are any number of dogs in the world, but there are precious few wild animals. So each one makes an important contribution to the species. So you have a responsibility not only to see it survive under your care, but to get it to breed."

But in the end, some animals have to be killed. That is the only way to stop the pain. Because the cellulitis was so far advanced, Cloris the slow loris had to be euthanized. Her tongue was paralyzed, her jaws didn't function. There was no hope. "The loris was a mess," said Leslie.

At the Penrose Annex, Keith picked the loris up in his arms, found a vein, and filled it with an overdose of ketamine. This is the same anesthetic the vets use for knockdowns; to euthanize an animal, they just use more of it. The animal becomes progressively more disoriented and finally goes out in a hallucinogenic blur. Cloris went limp almost instantly. In seconds she was asleep. A few seconds more and she was dead. Keith felt for a heartbeat and finding none, carried her in his arms to the pathology lab in Penrose. He gently laid her on a wooden shelf inside the walk-in freezer to await a necropsy by Dr. Robert Snyder, the zoo pathologist. Then he shut the door and returned to his work.

There are two ways to administer a lethal dosage, Leslie says; one of them is in the muscle tissue, the other in the vein. The muscle tissue is a surer shot, since there is greater mass to hit, but it takes a few minutes for the injection to work. With the vein, you run the risk of blood clots and the possibility of "blowing the vein" by accidentally popping it with the needle. But it is quick. "You press the needle in and they're gone," she says. Vets always go for the vein. "It's better because it's instant. Nobody likes the job, and they want to get it over with as quick as they can."

8

Toward the end of October, Dave Wood finally cornered the stray cat that had spread toxoplasmosis around the kangaroo yard. It was a white feral cat, not much more than a kitten. Wood had been trying to trap the cat for three months. He'd spotted it around the kangaroos a couple of times, and once more out in back of the elephant house, but he hadn't been able to get to his gun in time to kill it. "We were really after that cat," he said.

He was coming back from a search for dead mallards around the edges of Bird Lake (any mallard corpses have to be pulled from the lake lest other animals feed on them and spread botulism and other diseases) when he spotted the cat in the kangaroo yard once more, and this time he had his rifle. "I looked at him," he said, "and he stared back at me." But Dave couldn't shoot because he was afraid that he might miss and hit one of the kangaroos. "I tried every human way to get the thing, but I couldn't get a clear shot," he said. Finally, the cat sneaked away.

The next night Dave had stayed up late to watch *Cheers* on the conference room TV with John Groves and Eileen Gallagher. They had left ahead of him. Dave had gotten into his car and flipped on the headlights to go home, too. And there, in the low beams, was the cat, standing next to a fiberglass rock by the toolshed outside the kangaroo exhibit. Dave had a .22 rifle in his office. He ran quietly to get it, drew the gun to his shoulder, and got the cat in his gunsight. He fired, and the animal spun around with the first bullet, landing on its side. He fired again, and it was still. He walked up to the cat and emptied the rest of the chamber in its head.

Dr. Snyder did the necropsy and confirmed that the cat was indeed carrying toxoplasmosis. Laboratory analysis revealed the tell-tale moon-shaped trophozoites in its tissues and blood. The corpse was in the freezer of the pathology lab now. I asked if I could see it. "It's just a dead cat," he said. When I persisted, Dave led me around to Dr. Snyder's laboratory and back to the freezer where Keith Hinshaw had placed the dead loris the week before. He swung open the heavy wooden door and took me inside. Lined with wooden shelves, the room was like a butcher's meat locker. The shelves were bare except for some spare equipment—some coiled hose, a few test tubes. The cat was in a plastic garbage bucket on the floor, hidden under the carcass of a pregnant iguana named Ophelia. The iguana had slipped off a rock in her exhibit, cracked open some of her eggs, and died of peritonitis. There was a snake underneath the cat. Dave didn't know what had happened to that one.

Curled up with the iguana and the snake, the cat looked like the subject of a still life. The body was cold and limp, and the normally black membranes around its nose and mouth had turned gray. But otherwise the animal might have been sleeping. I didn't notice any bullet holes, or blood. It was an uncommon end for a stray cat—shot to death at the zoo and dumped in a garbage can in the freezer. But Dave didn't make anything of it. He merely looked at the cat with a kind of oh-*there*-you-are expression, as if it were a lost sock. I asked him if he had had anything against the cat. He looked at me blankly.

"It had killed off two wallabies and two kangaroos, after all," I said.

"I don't blame the cat," he said, in a way that suggested it was a peculiar idea. "It wasn't *his* fault."

"Then how come you shot him six times?"

"I wanted to make sure he died quick."

* * *

Like the cat, all the animals at the zoo end up on Dr. Snyder's table eventually, to be dissected by his scalpel. Massa had been there, flat on his back, while Snyder cut open his belly; the baby rhino that had lived only one day; hippopotamuses, alligators, kangaroos.

"One thing about being a pathologist," said Snyder, "you learn how to sharpen knives." For the smaller, more delicate animals, he can get by using a scalpel. For the bigger animals, or for the rough stuff, he uses a butcher knife. Snyder, in fact, had done some butchering in a meat shop as a teenager. He learned some of his techniques there.

The hides of rhinos and hippos are predictably tough. Those of buffalo and giraffes are surprisingly so. "Giraffes really have the thick skins," he said. "It keeps the lions off." The skin was thickest, nearly an inch, on the giraffes' backsides; it thinned down on their fronts, where they could protect themselves better. Massa's skin was as thin and pliable as a human's, but his skull was so thick Snyder had had to cut through it with a hacksaw.

Snyder is one of only a half dozen zoo pathologists in the country. "All the big zoos have them," he said, "if they're any good. How else are you gonna know what's going on?" At fifty-nine, Snyder was possibly the dean of pathologists. At the Philadelphia Zoo he had started as a researcher specializing in primates, then taken on his long-term study of hepatitis B and liver cancer in woodchucks. He had always done pathology on the side. Of all the zoo staff, he was possibly the most learned, ready to give a disquisition on just about any aspect of natural history at a moment's notice. He could also be absentminded. Eileen Gallagher had taken to taping notes to his briefcase to make sure that he remembered his appointments.

In performing a necropsy, he found that it went best in pieces. He would lop off the legs first and drop them in a bushel basket, then pull out the organs and put them in another basket, saving the most interesting bits for pickling jars. Preferring the larger organs from the big animals, he had retained portions of Massa's brain, heart, and lungs as mementoes of the illustrious ape. He often hummed while he worked.

"Ever seen an elephant heart?" he asked.

I shook my head.

"It's impressive, really impressive. Must weigh twelve, fifteen pounds."

He had come across some strange things as he conducted his investigations. In one coscoroba swan—a large, white, graceful bird—he'd found assorted change and a flashlight battery. An alligator had swallowed a pair of glasses, a wristwatch, and a set of false teeth. "He must have found the stuff in the water, and gobbled it up," he said. "Our old hippo Jimmy had a penny caught between his teeth. It was almost completely worn down by the time he got to me." Elephants are, fortunately, fairly picky eaters, but Snyder has turned up a paper bag or two in their intestines. "That's why we have to keep the litter from blowing into the enclosures."

As he talked, he opened up a euthanized woodchuck from his collection. He flipped it on its back. He took his scalpel and ran it from the creature's throat down to its crotch, and opened up the fur as neatly as if he had found some invisible zipper. "The shoe industry uses the woodchuck pelts for shoelaces," he said. "It's tougher than you'd think."

He hacked at the rib cage, and pried it open. Woodchuck guts spilled out onto the stainless-steel tabletop: the long, rubbery coils of the intestines, the plumlike heart, the twin ears of the kidneys. He pulled out one of the kidneys to show me. "See?" he said. "Chronic infection." The kidney was generally the shade of sautéed mushrooms, but with his thumb he pointed to a tangle of red lines, like a bloodshot eye.

Normally, he said, the heart was enlarged as a result. It had to work harder, since the kidneys weren't doing their job of purifying the blood. He placed the heart on the scales. "Fourteen grams," he said. "Hmmf. That's not what I would call enlarged. I'd have expected fifteen grams at least."

He dropped the kidney on the scales: 13.8 grams. "Normally, it would be twenty-three," he said. "It's shrinking because of the disease. I guess the heart's just started to compensate. Oh, now look there"—he pointed to the ear canals, which were inflamed—"infection. He had chronic earaches, this guy. When his kidneys went, it wore out his immune system. He probably had a lot of colds, too.

"It's detective work, this part of it," he said. "It's fun." He sliced open the stomach and some shredded green stuff spilled out. "That was his last meal, right there," he said. "Some greens. Stomach looks okay, though." He poked around with his gloved fingers. "No ulcers."

"Where's the soul?" I asked.

"The what?"

"The soul."

"In the pineal gland," he said.

"In where?"

That brought a smile to Dr. Snyder's lips. "It's a third eye that reptiles used to have, a light receptor. It's rather mysterious. The pineal gland is all that is left of it. It's thought to be related to the animal's sensing of the changing of seasons, and it may help animals time reproduction. There is something like it in mammals and birds. Early anatomists thought it was the seat of the soul."

"Maybe it is," I said.

"Sure," he said carelessly. "Why not?"

He dropped some of the organs into pickling jars for further study. The room took on the rank smell of formaldehyde. He told me that, possibly owing to the gloominess of their trade, many pathologists had turned to alcohol. One of them, on a bender, had downed a glass of formaldehyde during one operation. It was unclear if he had mistaken the clear fluid for gin or water. "He got awfully sick," Snyder said slyly, "but the formaldehyde really fixed his esophagus."

The rest of the woodchuck corpse went into a bushel basket. Like all the dead zoo animals, except for Massa whose body was parceled out to labs around the country, it would ultimately go to a rendering plant to be reduced to protein and made into fertilizer. It would be returned to the soil.

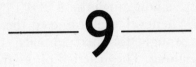

9

Up in the ADOPT office, off the marketing department in the administration building, Ann Novak was bothered by the death of Cloris the slow loris. Ann is a kindly, middle-aged woman with sad eyes and a quiet voice. She runs the ADOPT—for Animals Depend On People Too—program in which the public is invited to contribute money to support zoo animals, somewhat in the manner of the foster-child program for Third World relief. The idea, and the acronym, were taken from the Chicago's Lincoln Park Zoo in 1979. But so many ideas are filched in the zoo world that nobody minded.

It was not Cloris's death alone that disturbed Ann Novak, however. It was the fact that it had come so close on the heels of her cagemate Boris's demise. Ann Novak had just finished sending out form notices to the "parents" of Boris explaining, albeit somewhat circuitously, that he was "currently not on exhibit at the Philadelphia Zoo," and that perhaps they would like to transfer their donations, and their affection, to Cloris. And then Cloris died!

To complicate matters still further, Wilbur Amand had announced that a new slow loris was on its way to the zoo from the breeding colony at Duke University. But it was male. If it had been female, Ann was not above labeling it Cloris and

205

letting it go at that. But it wasn't. So what was she to do? Say that Boris had come back? Pass the new male loris off as a female? She had already heard from some of the Boris people. A middle-aged woman had called to say that her nephew had been "very upset" to hear of Boris's departure. "I'm in a quandary," said Ann.

It was impressive how much the ADOPT "parents" felt for their "children." Whenever they came to the zoo, which was often, they made straight for their animals' cages to see how they were coming along. They carried photographs of them in their wallets. Some major contributors displayed eight-by-ten photographs on their mantelpieces. And when the "parents" were away, they sent letters. Ann had one from a couple urging her to give their turtle, Myrtle, a pat on her shell. Some "parents" wrote to congratulate the animals for special accomplishments. A pair of elderly ladies had jointly adopted the rhino Billy. When he succeeded in impregnating Xavira the second time, they sent him a card. "Attaboy!" it said. One woman sent greeting cards to her aardvark at Christmas, Valentine's Day, and Easter, enclosing each time a small check, and signed the cards, "Love, Mom." She wrote one note from a grocery store in California where she had gone to get the aardvark a snack of some termites. She was sorry to report the store had none in stock. When the aardvark had to be sent off to San Diego to make room for her daughter, Aava, the woman magnanimously agreed to adopt Aava, and then offered to send Aava to San Diego, too, so that mother and daughter could be together. The zoo declined the offer. A doctor and his wife had adopted a small screech owl. They were very fond of it, and displayed pictures of it all around the house. Then the screech owl died, and the zoo sent them the usual letter about the animal not being on exhibit anymore. The couple called to ask what the letter meant. Ann had to tell them that their screech owl was dead. There was silence on the other end of their line. Ann said there was another screech owl coming in, would they like to support it? The couple said they would have to think about it. A week later, Ann called again: What were their plans? The couple said that their hearts had gone out to their original screech owl. They couldn't switch now.

In general, the public supported the usual animals—the

big cats, elephants, bears. Reptiles were not in much demand except for an occasional frog or turtle. Birds were problematic because they were generally so short-lived. The zoo didn't like to push them. But there were special cases: the Baird family, which adopted a bird, as did some people who lived in Bluebird Lane. (In the same vein, one family selected a fox in the name of their dog. Another family, not to be outdone, adopted some mice in the name of their cat.)

A key development in the ADOPT program had been the increasing use of individual names for the animals at the zoo. Names had always posed a philosophical problem for some of the relatively few hard-liners on the staff who hated to see animals anthropomorphized by being called Webster, Frankie, John. . . . As a result, names were rarely posted on the signs identifying the residents of various cages. It was considered demeaning, and it ran the risk that the public might overly identify with one individual and be unnecessarily disturbed if that animal wasn't there one day. As a rule, the descriptive signs were limited to species name, natural habitat, and a few behavioral characteristics. Nevertheless, most of the larger animals had names, and those names were used regularly by the keepers. Occasionally, the administration tried to steer a middle course by giving an animal a name that was redolent with associations of his native habitat, and therefore carried a reassuring scientific aura about it, but those names were often unpronounceable. A male giraffe, for instance, was officially named Rafiki Kodogo, which is Swahili for "little friend." But he was known to all at the zoo as Dinky. By the same token, the name Kanakbala was simplified to Billy. On the other hand, the names did provide a useful handle for individual animals.

The ADOPT program didn't catch on until it gave names to the animals it was marketing. Ann Novak herself had pressed to get the name of Boris for the late slow loris. That sold him. And now look what happened.

Ann herself knew what it was to care about an animal. She had worked for sixteen years in the Children's Zoo, and developed many attachments along the way. It was she whom the yellow-headed Amazon parrot had been calling when it shrieked "Yo! Ann!" so confusingly at Dave Wood's arrival. "I've shed my tears over animals down there," she said. She was thinking of one chimpanzee in particular whom she used to take by the hand from his overnight indoor cage to his outdoor cage every morning. Finally, after several years she switched jobs. The chimp never got over it. The first time the new keeper came to take him, he broke free from the man's grasp and ran off to look for Ann. She had to hide behind the refreshment stand. From there, she watched him search all over for her in vain. That broke her heart. For weeks afterward, every time the chimp was led to his cage, he would turn his head this way and that, trying to find Ann. And there was a parrot that had learned to bark like a dog and give the raspberry. She had to send him to the governor of Bermuda as a special gift from a zoo. That was hard. But the worst time by far was when she had to send away a handsome gray fox. The fox behaved like her own pet puppy dog. He didn't so much as look at anyone else, but when Ann called, he came bounding over and jumped into her lap. Finally, the zoo sent him away to the New Orleans Zoo. Since she was the only one he trusted, they asked her to do the crating. "That was the hardest thing I've ever done, " she said. "I put him in the crate, and finished nailing him up. I can still see his eyes looking up at me as I nailed down the lid."

She had decided what to do about Cloris, she said. "We're not going to let anybody know that there's a male in the cage. We're just going to let everybody think it's a female." She wasn't happy about the decision. But this way the "Cloris people," as she called them, wouldn't feel so bad, and the animals would go on being provided for. Before too long, the zoo would probably get another female slow loris from Duke, and everything would be back to normal. It wasn't a good situation, but for now it was the best she could do.

—10—

November loomed and still there was no sign of Xavira's impending delivery. I often hung around the fence by the rhino yard to check on her and see if I could spot any changes in her behavior. Was she getting cranky? Fidgety? There were a couple of times when she became irritable. In the middle of the month, she shook off a bird that had settled on her shoulders. That was widely reported. A week later, she chased Chuck Ripka out of her exhibit the way an angry bull goes after a matador. He came in all out of breath to report that to Dave Wood, but Dave didn't think it meant so much. And sure enough, Xavira calmed down again right after. She ate her hay, made her dumps in the same spot, and generally loped around the yard as though she hadn't a care in the world.

It was Billy who had changed, I thought. He was two stalls down from his best beloved. The zoo didn't want him to distract her in her delicate condition. Besides, male and female rhinoceroses have little contact in the wild; they come together only to breed. A couple of Malayan tapirs—porcine black-and-white creatures called "blanket" tapirs because a patch of white covers their backsides like a cloth—occupied the middle stall between the two rhinos. The tapirs had been separated once little Denise came along because the father had attacked his daughter. Now that Denise was grown, she had been shipped out, and the breeding couple happily reunited in their old paddock.

Occasionally, Billy would cast loving glances at Xavira. He would hang by the stone wall lining one side of his enclosure and, in a lovelorn way, rest his massive reptilian-looking head on top of the wall and moon at her like a lovestruck teenager. Then he would snap to, and trot off again.

Their romance would renew a species that, according to fossil records, had been going for sixty million years. It was descended, in turn, from an ancestor named *Baluchitherium grangeri* that went back even more millions of years and was probably the largest mammal of all time, standing nearly twenty feet at the shoulder and weighing twenty-five tons.

Despite its age and some design flaws, the 60 million B.C. model still worked remarkably well. Take Billy, now chewing hay in his paddock, a line of drool hanging out of his mouth. His armored skin looked like a sweater that had stretched in the wash. He had bunched it up in places to restore the fit. He had a ridiculous pom-pom of a tail, and a silly but endearing bit of fluff at the tips of his fat donkey ears. And up in the prow, he possessed that famous horn. He'd worn his down to a nub from years of scratching it against the wall, rubbing it on the ground, and absent-mindedly bumping it into things. Rhinos see very poorly. Their eyes are spaced so far around the sides of their heads, they have to swing this way and that in order to take in what is directly in front of them. And they notice only moving objects; stationary objects disappear into the background. Yet what rhinos lose in eyesight, they gain in smell. Their olfactory nerves take up a larger part of their bodies than their brains.

Nevertheless, the bulky rhino is most likely the source of the unicorn myth. In the Linnaean taxonomic system, the Indian rhino is called *Rhinoceros unicornis*. In the account of his travels to the Far East, Marco Polo called rhinoceroses "unicorns," meaning "one horns," when he spotted some in Sumatra in 1298. But he noted with evident disappointment that this unicorn was "altogether different from what we fancy."

One part of the male rhino, however, has never disappointed—the penis. Billy was once pictured in *Hustler* magazine in full erection: It looks as if he is straddling a rocket ship. The photograph held a prominent place near Billy's indoor stall in the elephant house. This photo was generally considered more interesting than Billy's other media appearance—in a print advertisement for "rhino tough" Armstrong tiles, in which Armstrong had tiled the floor of his stall and posed Billy standing there, trying to look tough, but actually appearing, to those who knew him, a little bewildered.

When fully extended, Billy's erection was somewhere between two and a half and four feet long. No one had ever put a tape measure to it. Generally, when zoo people were asked about it, they would stretch out their arms about as wide as they could go, and say, "Oh, about that long." His penis had interesting little flaps on either side, like stubby airplane wings, to lock it in place when coitus was achieved. Chuck Ripka said it looked to him "like a goddam French tickler." Billy often aired it out a bit in the mornings, although he rarely displayed the full regalia; it hung down between his legs like the proverbial hose, and caused at least one small child to ask his mother why the rhino had five legs. According to Chuck, some of the women visitors were even more entranced. "They see him get this thing and they go, 'Ohhhh,'" he said. "Then they ask me, 'What do you feed him?'"

Dr. Steven Jaffe, one of Keith Hinshaw's predecessors, had once considered obtaining a semen sample from Billy. The normal procedure is to administer "deckbrush anesthesia"— masturbating the animal with a long-handled scrub brush. A big, powerful animal like a rhino requires a different method. Dr. Jaffe thought that he would climb inside a crate covered with mattresses, have the crate placed in Billy's enclosure, wait for Billy to mount it, then somehow use his gloved hand to bring Billy off. "Vets don't mind that kind of thing," said Dave Wood. "But you wouldn't catch me doin' it." It was an unusual concept. Jaffe got the crate, tacked the mattress down on it, and hauled it to the fence, where he explained his plan to Jimmy McNellis, the man known around the elephant house as God. God stood there with his arms folded across his chest and looked very dubious. He had heard a lot of crackpot ideas in his time, he told Jaffe, but this one had to be the worst.

To humor Jaffe, he suggested that he try a small experiment first: Why not place the crate in the exhibit all by itself and see what happened? Jaffe thought that made some sense. So a couple of workers planted the crate in the exhibit, and they let Billy in. Billy took one look at the strange obstacle in the middle of his yard, and he let out a snort. He put his head down, charged forward, scooped up the crate with the tip of his horn, and smashed it against the wall, sending pieces flying in all directions. He butted the crate again and again in increasing fury. Finally, when it was reduced to kindling, Billy flicked the remains over the fence and into the dry moat. Dr. Jaffe went pale watching this spectacle. When it was over, and Billy trotted back around his yard looking for something else to do, Dr. Jaffe turned and put his arm around McNellis. "Thanks," he said.

Because of its impressive size, the rhinoceros penis has assumed a lofty position in primitive folklore, much to the rhinoceros's hazard. In imitation of the Sumatran rhino (a hair-covered version of the Indian, now extremely rare), some Borneo tribesmen insert a cross-strip of bamboo, bone, or wood into their own penises. And their tribeswomen hang a rhino's tail over their beds when they go into labor in the belief that it will ease the pain of childbirth. If the pain continues to be harsh, nursemaids hold a rhino's penis (detached from the rhino) over the woman's head and pour water through it.

Other cultures have focused on the rhino's horn as the source of procreative powers. In powdered form, the horn is marketed as an aphrodisiac in Singapore, selling for as much as $30,000 a pound, although the potion is clinically speaking no more worthwhile than biting your fingernails, since both consist of keratin. (Contrary to appearance, the horn has no bony core; it is all densely matted fiber.) Elaborately carved, the rhino horn is also sold as a ceremonial dagger called a *jambia* to oil sheikhs in North Yemen at equally outlandish prices, up to $12,000 apiece. Such markets have imperiled the rhino in recent years. According to David Western, a Kenyan wildlife ecologist, in 1985 the number of Javan rhinos was down to 65 animals, the Sumatran to 500, and the Indian to 1,700. Although the African black and African white rhinos number 8,000 and 3,000 respectively, Western wasn't confident that they were doing very well even at those numbers, since they were rapidly losing their native ranges.

At the zoo, therefore, it was of some urgency that Billy put his member to good use. The problem was, he didn't know the first thing about it. Unlike most zoo animals, Billy had actually been born in the wild, but he had been brought into captivity long before he had a chance to observe his elders' sexual techniques. Until Xavira came along, Billy hadn't had anybody to practice with either. For many years, Billy's sole consort was an aged female named Golden Girl who wasn't the least bit interested in sex. When the zoo let the two of them cohabit, he would virtually beg her for it, but she wouldn't yield. Ultimately, Golden Girl was found to be infertile; she never came into heat.

As the years went by, Billy grew increasingly desperate for some action. Because of Golden Girl's disinclination, Bill Donaldson says Billy's keepers used to go out and whack the rhino's immense penis with a shovel to discourage his ador. Out of sympathy for poor Billy and a keen interest in some Indian rhino progeny, in 1979 Wilbur Amand, as temporary president of the zoo before Bill Donaldson, sent to the Basel Zoo for a luscious, three-ton morsel of rhino flesh named Xavira— Billy's Swiss Miss. In a package deal, she and her brother, Assam, were flown to the United States on Lufthansa Airlines in a pair of large crates.

Named after the "Happy Hooker" (despite the difference in spelling), Xavira was undeniably a knockout. But Billy's painful memories of the shovel treatment proved even stronger than her sex appeal. When Xavira came into estrus (the keepers could tell because she was jumpy and she wasn't eating), they moved brother Assam indoors out of sight and let Billy in with her for a date. Billy took one look at her and went back to eating hay. His keepers worried that Billy had taken leave of his senses. "He had been mooing at the moon for years," said one staffer, "and then when he has his first big chance— nothing."

Forty-five days later, Xavira was acting frisky again. This time, when she squirted urine in heat, Billy, in his adjoining pen, pulled his front lip back in a peculiar gesture, called the flehmen response, that exposes some delicate sensors inside the upper lip and makes the rhino look like Mr. Ed talking. It's a rhino's way of expressing interest. In his eagerness, Billy started butting the gatepoles, trying to get at Xavira. The keepers let him into her pen at midday.

Sensing that something was about to happen, a number of keepers, staff, and visitors—some fifty people altogether— gathered along the fence to watch the scene. One of the visitors set his movie camera rolling. But as before, Billy spurned Xavira once he was in her yard. He trotted off into the corner by himself with barely a glance at her. This time, Xavira wouldn't take no for an answer, and she tried some of her feminine wiles. Rather than leave the first move to Billy, she sidled up to him and daintily rubbed her horn up against his cheek. Still nothing. Then Xavira ran her snout down the length of his body and tenderly began to lick his underparts. That registered. Billy let out a low moan, and his penis began to lengthen. Xavira, suddenly unnerved by what she had unleashed, backed off to the far fence. Meanwhile, Billy started to prance lustily. His erection was two feet long and growing. "It was incredible," said Janet Pensiero from the graphic arts department. "It went from his groin practically to the tip of his horn!"

Billy charged at Xavira, climbed up on her backside, and rammed himself up against her. But he missed! His aim was about a foot low. In his excitement, he ejaculated anyway. Like some lovestruck virgin, he was full of desire but completely ignorant of sexual mechanics. Watching from the sidelines, Dave Wood was distressed. He saw a great chance to increase the world's Indian rhino population suddenly being lost because of Billy's ineptitude. "I wanted to run in there, grab Billy's dong and go 'Here!'" he said with a gesture of his hands as though he was handling a fire hose. Billy's erection wilted and he slumped off her. Xavira curled up on the ground. It looked like Not Tonight, Dear.

But in fact Xavira was merely recharging her batteries. For then she got up and nuzzled his tender parts again. And Billy responded eagerly. This time, Xavira climbed down into the dry moat, and leaned suggestively against the wall. Unfortunately, that left Billy no way to get on top of her. Dave Wood couldn't bear it. He hopped over the fence and climbed into the moat to reposition her with a few pats on her rump. But then just when everything was ready, Xavira coquettishly climbed back out of the moat. Billy lumbered after her. They bumped heads for a moment or two, then Billy swung around to get up on her, but his aim still wasn't right. He slid off. Xavira returned to the moat. Once again, Billy followed her. This time, Billy mounted her smoothly, his erection growing as he ascended. The onlookers heard an odd slapping sound as he tried to pop himself into her, and they could see his haunches heave desperately. Finally, there was a delicious silence. He'd thrust himself in—and in and in. Xavira gave out a gasp, and then the two bellowed together, as Billy slid himself in and out on top of her. With each thrust, his head bobbed like a jazz musician hearing a really good sound. He ejaculated quickly. The onlookers could tell because Billy's knees buckled and his little tail shot straight up into the air. But that was just for starters. He climaxed again and again—every fifty-five seconds for the next hour and a half. When he was finally spent, Billy slid off and lay down, panting. Xavira shook herself, and some semen gushed out of her vagina. She curled up next to him, her head on his belly. The two looked very much in love.

When they had recovered, Billy was led away back to his paddock, Assam was returned to his sister, and the onlookers dispersed. "I don't know about the others," said Janet Pensiero, "but I had to go home and take a cold shower."

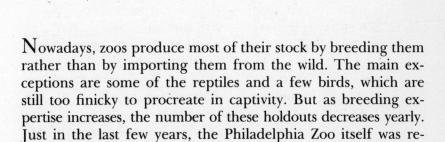

Nowadays, zoos produce most of their stock by breeding them rather than by importing them from the wild. The main exceptions are some of the reptiles and a few birds, which are still too finicky to procreate in captivity. But as breeding expertise increases, the number of these holdouts decreases yearly. Just in the last few years, the Philadelphia Zoo itself was responsible for the first zoo births of the diademed tanager, hooded pitta, and several others.

Indeed, the zoo's very success in captive breeding has caused some concern that zoo animals are slowly evolving into different species from their brethren in the wild. This occurs less in terms of the animals' physiognomy—zoo leopards still have their spots—than in terms of their behavior. It is probably inevitable that zoo animals lose some of their predatory instincts in an environment where dinner is handed to them, but this process is being speeded at less sophisticated zoos than the Philadelphia that prefer animals to be easy to manage rather than to be true to their species. Zoo animals, in short, are being domesticated.

This wasn't possible until fairly recently. In the old days, when a zoo needed a new animal it would call up a trapper, who would then dash off into the jungle to fill the order. In

1922 when the Philadelphia Zoo wished to replace old Pete, the rhino it had received from P. T. Barnum, Charles Penrose contacted Frank Buck. By far the best known animal collector of his day, Buck had begun his career as the head of publicity (among other duties) for the Western Vaudeville Managers Association and, truth be told, he never lost his knack for promotion. Invariably attired in his jungle khakis, and tropical hat, Buck was once described as the country's oldest Boy Scout. But he did catch a lot of wild animals. By his own count, Buck captured thirty-nine elephants, sixty tigers, fifty-eight leopards, fifty-two orangutans, five thousand monkeys, two giraffes (difficult to transport because of their height), eleven camels, ninety "large" pythons (meaning more than twenty feet long), and fifteen crocodiles. In the process, he once knocked out an orangutan with his bare fist, narrowly escaped death from the jaws of a frenzied tapir, and singlehandedly captured a leopard that had escaped from its cage on a passenger steamer. His misadventures, however, reveal the Boy Scout in him. He nearly froze to death one time in a ship's refrigerator. He had gone to get some chopped liver for his pheasants when someone accidentally locked the door on him. And another time he nearly died while plucking some dead skin off the head of a king cobra. Some of the cobra venom had dribbled onto his hand; a fly bit him on his moustache; he scratched the bite, inadvertently rubbing the venom into open cuts left on his cheek from shaving. Dashing to his medicine cabinet, he rubbed some lye into the cuts to stop the poison before it reached his bloodstream.

Frank Buck told of capturing the rhino that came to be called Peggy in *Bring 'Em Back Alive,* his first and most famous book of adventures. The trip came about at the behest of William Temple Hornaday, the illustrious head of the Bronx Zoo, who first asked him for an Indian rhino; the Philadelphia's Charles Penrose got in on the venture later. The joint request was a little like asking for a moon rock. It was out there—just hard to bring back. Even in those days, before poaching so drastically reduced their numbers, Indian rhinos could not just be plucked off the rice paddies. The greatest number of them were in Nepal, but that country offered two formidable problems. The reigning maharajah regarded the rhinos as his private reserve, and no foreigners were allowed into the country. "I like tough jobs," wrote Buck gallantly. After lining up other requests for specimens, he set sail for Nepal, by way of Hong Kong and Calcutta, in May 1922.

In Calcutta, Buck made the sort of contact that is invaluable for men of his profession. He met General Kaiser Shumshere, a nephew of the Nepalese maharajah, who had taken up temporary residence in the city. Shumshere had set up a rhino hunt for the prince of Wales the year before. With some bluster from his interpreter, Buck managed to get by some well-armed and imperious sentinels from Gurkha—a Nepalese state famed for its warriors—and into the sumptuous private apartment of the general. Shumshere was dressed in jodhpurs of pink silk, a vest of green velvet, and a white silk shirt. Despite his princely garb, he had a businessman's approach to life and, seeing that there was money to be made, agreed to telegraph Buck's request for two rhinos to his uncle the maharajah. Four days later, Buck was summoned back into Shumshere's presence. The price was 35,000 rupees, or $12,600. Buck told Shumshere that was a lot of money. (It certainly was—but no more than the $80,000 apiece they would be worth today.) Shumshere passed Buck's reply back to his uncle, but that was as low as the ruler would go. Frank Buck reluctantly said he had a deal.

Shumshere himself would head the expedition to collect the rhinos. Collectors generally delegated that messy job to the locals. Traders didn't know the territory as well as the natives, and they would just as soon let the outraged animal take out its frustrations on someone else. Their role was to be the diplomats, the bankers, the organizers, and, once the animal was obtained, the travel agents and escorts for the long journey home. Frank Buck waited in Calcutta while Shumshere went into the rice paddies across the Nepal border to hunt the rhinos.

To be sure, the art of animal trapping did have a few tricks that reduced the risks. In a 1922 book with the wonderful title of *Trapping Wild Animals in Malay Jungles*, collector Charles Mayer described using birdlime, a sticky substance, to immobilize tigers and leopards. "Immediately after a cat animal has put his foot in the stuff, he becomes so enraged and helpless that he is easily captured," he wrote. "The tiger or leopard that steps in birdlime doesn't step gracefully out of it and run away; he tries to bite the stuff from his feet and then he gets it on his face. When he tries to rub it off, he plasters it over his eyes. Finally, when he is thoroughly covered with it, he is so helpless that without much danger he can be put into a cage; and there he spends weeks in working patiently to remove the gum from his fur." Birds and monkeys also succumb to the birdlime treatment. For small monkeys, though, Mayer stuffed a sweet-smelling rag into a bottle and tied the bottle to a tree. Apparently, the monkeys poked their hands in to grab the rag, but couldn't bear to let go of the rag in order to pull their hands back out. "There he sticks, fighting with the bottle, until the hunter comes along, and, by pressing the nerves in his elbow, forces him to open his hand and leave the rag for the next monkey."

Other techniques were more cruel. With herd animals, it was possible to run the herd until the infants lagged behind, exhausted and ready to be scooped up by the collector. That worked with giraffes, zebras, antelope, and, in one instance, rare Przhevalski's horses from Mongolia, which the noted German collector and zoo builder Carl Hagenbeck had been hired to secure for the duke of Bedford. More daringly, the Havati people of North Africa swam up to young hippopotamuses and even crocodiles, and harpooned them—deeply enough to disable but not so deeply as to kill.

Or collectors used traps. For smaller animals, these generally consisted of a noose that closed around a leg and whisked the animals up into the air. For larger beasts, a pitfall was dug into the ground and covered over with brush. The animals came lumbering along and tumbled in. Young hippos were particularly vulnerable, for the mothers customarily drove their progeny in front of them. In his memoirs Hagenbeck wrote that when the hippo mothers saw their children vanish into the pit, they abandoned them and ran for their lives.

But generally with sizable mammals, the trappers used the time-honored method of killing the mothers and grabbing the helpless infants. Hagenbeck wrote respectfully of the Nubian swordsmen who killed mother elephants by swooping past them on fiery Abyssinian ponies, slicing their leg tendons with swords, and leaving them to bleed to death while they made off with their young. Usually mother elephants were shot by rifles from a safe distance. This approach worked also for African lions, gorillas, polar bears, and Indian rhinos.

Being businesslike, General Shumshere chose that method himself. He organized a private army of a hundred formidable Gurkhas and thirty elephants, and went out to find nursing rhinoceros females. Finally, he surrounded two mothers and their calves, and dropped them with rifle shots. Pathetically, a calf who is suddenly orphaned in this manner does not flee but will hang by its dead mother for days until she begins to decompose. The Gurkhas had no trouble constructing a log corral to fence the baby rhinos in. When that was done, they extracted their mothers. Shumshere had Buck's rhinos.

As soon as Frank Buck received word of the expedition's success, he took a train from Calcutta with the payment of 35,000 rupees, in cash, bound up in a money belt that he slung around his neck. Unfortunately, only cash would be accepted. Halfway along, however, he discovered that the Ganges River had overflowed its banks greatly impeding further travel. He telegraphed the men holding the rhinos, and remarkably they agreed to hold on to them while he dashed back to America with a shipment of animals that had been left waiting in Calcutta.

Several weeks later, he returned with his money belt and went by train as far as Raxaul, a small Indian city on the Nepalese border. From there, he traveled by mountain pony across the Nepalese mountains to Bilgange, thence by elephant to the camp with the rhinos. By then, General Shumshere had returned to Calcutta. At the camp, the Napalese official in charge wanted full payment immediately. Buck much preferred to make payment when the animals had reached the train station at Raxaul across the mountains. To see that he obtained his preference, Frank Buck pulled his gun. "I had no desire to fight the Gurkha army," he wrote, "but I was perfectly willing to plug a few of these boys if they forced a scrap." Even more effective than the way he brandished his revolver, however, was Buck's assertion of close friendship with the revered General Shumshere. When Shumshere's name was invoked, the official declared he would be delighted to receive payment wherever Mr. Buck wished.

When the procession finally started back across the mountains the way Buck had come, it made quite a spectacle. The baby rhinos traveled in two crates loaded on creaky wooden carts drawn by four water buffalo. Fifty mud-splattered Gurkha soldiers strode along with them. Then came three elephants, each piled eight feet high with leaves of the jackfruit tree to feed the rhinos during the journey. Buck and the official rode on ponies, and galloped back and forth trying to keep everything in order. They made it to Raxaul in three days. The train was a day late, but when it finally pulled in, the rhinos were loaded onto flatbed cars, their crates festooned with leaves to keep out the burning sun. There, Buck finally paid off the official. On to Calcutta!

The journey was marred, however, when a Hindu slipped past the guard one night as the trains stopped at a station. He made off with a small but infinitely valuable chunk of one of the rhinos' horns. Otherwise, all went smoothly. Once in Calcutta, Buck couldn't find a shed big enough to store his rhinos, so he left them on a friend's front lawn.

The Buck brigade, now swollen with other animals he was importing, set sail for America a week later. The ship encountered heavy weather in the Bay of Bengal. Just past the Philippines, it hit a typhoon. For thirty hours, the storm crashed and howled about the ship. The elephants bellowed and shrieked as the waves smashed over them. They tried to break the force of the waves with their heads. Through the torrent, Buck could see that the ropes holding the rhino crates fast to the deck were working loose. Despite the captain's orders, he slid along the foredeck, clinging to a wire cable to keep from being swept overboard, and crept out to the rhino crates. The animals groaned as they sloshed about in their cages. The ropes still held, but Buck was sure they would go any moment. For an hour, he labored to lash the crates tight once more, hugging the mast for safety every time a wave crashed over him. Finally, the rhinos were secure. A few hours later, the typhoon cleared, and the ship made it to San Francisco without further trouble. From there, the rhinos were delivered by train to Philadelphia and New York. Hornaday and Dr. Penrose were overjoyed to see Buck pull up with their rhinos. True to his word, he'd brought 'em back alive.

Peggy, as the Philadelphia Zoo's rhino was called, went straight into the old Pachyderm House, where she aged gracefully. When she was fully grown and weighed three tons, one newspaper said she was "as cute as a ferryboat." After coming halfway around the globe, Peggy stayed in her yard until the present Pachyderm House was built in 1941, almost twenty years later. Workmen constructed a special hundred-foot walkway to channel the animals into their new quarters. Most of them went amicably, but Peggy refused to budge until a keeper named Pat Cronin (who had earlier distinguished himself by yanking a dead tooth out of a hippo with a pair of pliers without benefit of anesthetic) cracked a whip behind her. That got her moseying along. A few yards down the runway, Peggy had an inspiration. She let out a roar and broke into a gallop. The bystanders and attendants all dropped their jaws as they saw three tons of rhino take off. Then they ran for cover. Bodies flew every which way. When the dust settled, Peggy stood at the far end of the runway, wondering what all the commotion was about, but no doubt secretly pleased with herself. Pat Cronin was pictured in the newspapers giving Peggy a kiss full on the mouth when he went off to war a year later. Peggy died the next year, at twenty-two, before Cronin came back.

Nowadays, zoo animals are still transported around the world on a regular basis. In most cases, it takes a matter of hours, for they travel in the pressurized cargo holds of commercial airplanes. What causes the difficulty is the quantity of regulations that comes into play when animals across national boundaries.

When Xavira and her brother, Assam, traveled from the Basel Zoo, they were trucked to Frankfurt and then flown by Lufthansa to Kennedy Airport in New York. And there, due to an oversight, they might have sat for some time if it hadn't been for the timely intervention of a modern-day Frank Buck named Frederik J. Zeehandelaar. He is a short, white-haired, curmudgeonly fellow of sixty-seven who uses an elephant-head cane that was carved in Africa, and smokes foul-smelling cigars. He is not one to enjoy the spectacle of failure. His only serious mishap occurred years earlier when a ship carrying an Indian elephant for the opening of a Boston department store lost its rudder off Bermuda, causing the elephant to miss the store's debut.

Zeehandelaar is a new-style animal collector; which is to say that he spends most of his time on the telephone. He is one of three major dealers in the business. The others are the Hunt brothers of Detroit, who run a big, international, and heavily computerized (thus winning Zeehandelaar's scorn) outfit, and German Schultz, the son-in-law of the proprietor of Lindemann's Animal Farm in New York State's Catskill Mountains. Zeehandelaar is Dutch by birth and speaks with a thick accent ("truck" comes out "turk"); the name Zeehandelaar in fact means "sea trader" in Dutch. He came to America after the Second World War and started a career importing spices and exporting industrial chemicals. One day, a clove shipper in Madagascar offered to include some monkeys in his next shipment, and his career changed direction.

Just by chance, in November of 1979, Zeehandelaar was sitting in the office of a U.S. Customs broker who was talking by telephone to Wilbur Amand about some rhinos he was flying in from Frankfurt. That got Zeehandelaar's attention. "Rhinos were coming in!" he said later. "Rhinos! That's not like a carton of cigarettes, you know. That's like the coming of Halley's comet." Naturally, Zeehandelaar asked the Customs man a few questions about the pending shipment. Zeehandelaar was shocked to hear that the Philadelphia Zoo's preparations had been in this case somewhat lax. "There were no papers, no documentation, no insurance, no arrangements, no truck, no *nothing!*"

Zeehandelaar couldn't believe it. He had imported twenty-five animals for the Philadelphia Zoo, starting with an aardvark and including the orangutan Bong, and he didn't want the zoo to blow this one. The rhinos were not his business, but he felt some moral obligation to retrieve the situation. He rushed back to his office and got to work.

Zeehandelaar's office is in New Rochelle, about fifteen miles north of New York City. He says that as long as he is located within reasonable distance of a major airport, he can operate anywhere. He occupies the second floor of an ancient office building above the main headquarters of the Road Runner Taxi Company and Bernice's Stationery. There are no animals in sight except some sculpted monkeys adorning an ashtray, and a hairbrush in the shape of a duck. The only living animals that Zeehandelaar ever sees are the "alley cats," as he calls his pet domestics, which he keeps at home. The office is musty and antique except for a Teletype computer in one corner. One wall is filled with animal books: The ones on African game are on the right, those on elephants in the middle, and those on extinct species are on the left. Zeehandelaar deals only in mammals. "Reptiles never appealed to me," he explains with some insouciance. "And I don't care for birds either. They are too delicate, physically too delicate. And the bird business is too complicated. Diseases, and dying, and sexing. I don't want it. I stick to the larger mammals." He keeps pictures of his more unusual acquisitions in a large storeroom across the hall: shoebill storks, monkey-eating eagles, a thirty-six-inch baby elephant, maned wolves.

He hit the telex machine as soon as he pulled in. The rhinos were headed for the Frankfurt airport, but they lacked insurance for the journey to Philadelphia and a truck to take them there from New York's Kennedy Airport. Zeehandelaar had to negotiate for $130,000 worth of "mortality insurance" from Lloyd's of London to cover the zoo ($80,000 for the breeding female, $50,000 for her little brother). Working late into the night, he tracked down German Schultz, who doubled as an animal transporter as well as a trader, to collect the rhinos in his specially outfitted moving van, and he greased the animals' transit through Customs. It was fortunate that these animals were rhinos. If they had been birds, they might have been subject to a one-to-two-month quarantine at their point of entry. Rhinos are not known to carry any diseases that are communicable either to man or to agricultural animals, and are therefore not subject to stringent quarantine upon entry to the United States.

Fortunately, the zoo had prepared all the tedious importation paperwork in advance or Zeehandelaar would never have gotten everything done in time to greet the rhinos when they touched down. "Ninety-five percent of my time is consumed by paperwork," he says. "I spend an hour a day on the *Federal Register,* just to see what proposals and what changes have been made." He deals with the Departments of the Interior, Agriculture, and Public Health. "They don't stand in my way, exactly. But they work for the government. They don't work for me. They are extremely slow. An endangered-species permit can take three months. The animals aren't any problem. It's people who are the problem." If the rhinos had come from the wild, Zeehandelaar's effort would have been impossible. He almost never imports animals from the wild anymore, with the occasional exception of bongoes from Kenya or okapis from Zaire, since they are still fairly plentiful. Most other wild species, especially an endangered one like the Indian rhino, are protected by their host governments because of their scarcity.

For his efforts, Zeehandelaar charged the zoo an extremely modest $50, plus telex expenses. In lieu of greater payment, he couldn't resist the opportunity to chastise Wilbur Amand for his handling of the affair. "It was not exactly easy to arrange this insurance on such incredibly short notice," he noted in a letter to Wilbur. "Only fast telex-work (partly during this night) saved this situation (plus—if I may say so—my relations)." And he closed: "Finally, I believe all went well?"

It did indeed. German Schultz arrived in time to greet the Lufthansa flight, the rhinos were trucked to Philadelphia, and promptly installed in the elephant house. Xavira was made ready for her introduction to Billy.

Since Billy had been caught in the rice paddies of the Indian state of Assam, for breeding purposes it didn't much matter with whom he mated. There was a safe assumption that he and his partner were not related. This is not always the case. As the number of zoo-bred animals increases, the hazards of in-breeding mount. The Detroit Zoo is a world supplier of polar bears. There isn't a polar bear in Australia's zoos that did not come from the Detroit. The Minnesota Zoo similarly specializes in Siberian tigers. With so many animals coming from such a narrow source, zoo directors have to be particularly careful. Until as recently as ten years ago, zoos kept poor records of their animals' genealogy, and they rarely shared their information with others. When zoos acquired another animal on a breeding loan, they took their chances. Breeding animals was not so important. There were always more animals in the wild if any one line died out. Besides, the zoos weren't particularly interested in sharing. They wanted animals that other zoos didn't have. This was the age of the "postage stamp collection." Directors wanted one of each.

Even today, this attitude prevails among directors. Bill Donaldson is a little surprised by his brothers in the ranks, many of whom display acute collectors' instincts. One of them collected state license plates. "He had every one ever issued," Bill marvels. "Must have had a thousand of 'em—government plates, press plates, handicapped plates. Another guy collected miniature railroad cars, dirty limericks—now they were something—and cap badges, you know the ones that go over the brim on baseball caps? I had one from the streetcar operator, and he wanted it, but I wouldn't give it to him, and he was furious." And he believes this collecting mania carries over into their desire to have certain rare animals, just because they are rare, no matter how expensive they are. "My friend Ed Maruska at the Cincinnati Zoo was going to buy an okapi. Gorgeous-looking animal, but it cost four hundred thousand dollars. I said, 'What if it croaks on ya?' I could put up two buildings for that. But that's what he wanted."

Yet in the last few years, a zoological organization has formed that has compelled the world's zoos to work together in a way that not so long ago would have been unimaginable. As part of an international program first proposed by a University of Minnesota researcher named Ulysses S. Seal in the early 1970s, over fifty-three thousand animals from 175 zoos and primate centers, including many (although not all) of the Philadelphia Zoo's collection, have come to be regarded less as their zoo owners' possessions, and more as resources for all zoos collectively. As part of the program, each of these animals possesses the equivalent of a social security number. In the case of mammals, the number is tattooed in India ink inside one thigh. It is called an International Species Inventory System, or ISIS, number (a wordplay on Isis, the Greek goddess of fertility), and it allows central computers to keep track of sex, age, parentage, and other biological factors that in turn allow zoo authorities to keep tabs on their animals' genes. As one official once explained, ISIS combines the duties of "the U.S. Census Bureau, the American Kennel Club, and a genetics counselor."

At Philadelphia, Beth Bahner is in charge of the record-keeping, which she manages from a small desk, surrounded by a range of file cabinets, in an office upstairs at Penrose. "It was a big job to get started," she said, "because we had to figure out the coding for all the species, and develop the forms and set up computer programs to keep track of everything. But it's been really valuable. Without these ISIS reports, we wouldn't have any idea what anybody else has." When the Philadelphia Zoo began its participation in 1975, it started with the major animals, then worked its way down. The zoo started with the primates, and happened to pick a Sumatran orangutan named Guas, the father of Bong's mate, Christine. He was MA (for mammal) 1000000. His card now has a line through it because Guas is dead. Ten years later, in 1985, the zoo was just beginning to record the reptiles. "We'll start with the turtles, then the crocodilians, and continue on in taxonomic order," said Beth. "It's going to take long time, and I'm not looking forward to it."

Dubbed an international dating service for animals, the ISIS system allows participating zoos to come up with good genetic matches for their breeding animals. "It used to be that if an animal bred, great," said Beth. "It didn't matter who they were. Now, these things are planned."

The most elaborate ISIS program is the Species Survival Plan (SSP), by which scarce species are singled out for special attention. There are currently about three dozen species in the program. In these cases, breeding decisions are actually taken out of the hands of the individual zoos and given to a special committee that coordinates the breeding activity on an international basis. Przhevalski's horses, the bushy-maned wild horses from Mongolia that Carl Hagenbeck had rounded up in the twenties, are on the list, since only a few hundred are left in the wild, as are all kinds of rhinos, Asiatic lions, orangutans, giant pandas, snow leopards, and many others. Larry Shelton, the zoo's curator of birds, administers the SSP for the Bali mynah, a large white bird with a blue swatch over its eyes, which used in fact to be called Rothschild's mynah after its Western discoverer. Selection is admittedly tilted toward the larger mammals, not just because they are more popular (although that doesn't hurt), but because they are, as Tom Foose, conservation coordinator for the AAZPA, once put it, more "critical. There's always going to be a good representation of beetles around for some time. But not such a good representation of rhinos."

But the best example, largely because it is so far advanced, is the case of the Siberian tiger. The SSP program is administered by Ulysses S. Seal, the founder of ISIS. As head of the committee, Seal decides which tigers breed with which, and when. The SSP is determined to avoid a kind of Siberian tiger baby boom, which would exceed the world's zoos' limited capacity for housing them. Zoo space is at a premium. William Conway, director of the Bronx Zoo, once calculated that all the country's zoos could fit into the borough of Brooklyn, New York. Room for tigers, of course, is a mere fraction of that, and it is full up. In 1980, when the Minnesota Zoo was trying to place a litter of three cubs, it could only locate space for one—in the Dallas Zoo. Besides seeking to maintain a steady population, the SSP would also keep from ultimately producing a surfeit of geriatric animals that were incapable of reproducing. And it wanted to avoid the kind of situation that nearly closed the Detroit Zoo when it needed to euthanize some aged and sickly Siberian tigers because it no longer had room for them. Word got out about the zoo's plans, and a citizen named Krescentia M. Doppleberger sued the city and the zoo for $1 million "for breach of trust, intentional inflicting of emotional stress and negligence." She lost but not before she had frightened the zoo administration severely.

This fall the Philadelphia Zoo was awaiting word on when its massive Siberian, George, would be allowed in with one of his three first ladies, Martha, Abigail, and Dolly. It was a little like waiting for airport controllers up in the tower to give a plane permission to land. Without it, the zoo, not to mention George, remained in a holding pattern.

Because Indian rhinos are so difficult to breed, Billy got the go-ahead as soon as Xavira arrived.

Originally, the SSP had hoped to go beyond merely maintaining captive populations to actually replacing depleted stocks in the wild. In the 1980s some successes have been achieved in returning the Arabian oryx to the Middle East, restocking the golden lion tamarin in Brazilian rain forests, and in restoring the nene goose in Hawaii. Researchers have also attempted various high-tech methods, including artificial insemination, cryogeny, and superovulation, to increase the dwindling populations. In 1980 a Holstein cow was the surrogate mother for a gaur, a large oxlike antelope. And later, a horse successfully delivered a zebra.

The successes of these programs have been limited, and they have been expensive. Consequently, zoos are rethinking their role in maintaining the supply of wildlife and blurring the distinction between the zoo and the wild. As native habitats have been severely reduced in size, intensive planning has gone into managing the resident species practically as if they were captive populations. At the Morris Animal Foundation for mountain gorillas in Zaire, for example, veterinarians have been introduced to oversee the care of the "wild" apes, which have suffered increased exposure to the diseases of tourists. The vets vaccinate the gorillas and monitor their health much as Keith Hinshaw and Mike Barrie look after John and his troop in Philadelphia. Zoo expertise, consequently, is in increasing demand at parks and reservations around the world—in the African plains, the South American rain forest, the Southeast Asian jungle. The world, it seems, is becoming a zoo.

—12—

Chuck Ripka is a relief keeper: He fills in at the various animal houses as needed. But he spends most of his time in the Pachyderm House, where for the last eight years he had been trying with precious little success to gain the respect of the elephants. Chuck is short and stocky, and his face, roughly bearded, is as ruddy as if he has a permanent case of windburn. He wears a tiny elephant pin in one ear. Add a red kerchief and some tattered clothes, and he would make a fine pirate.

With Gene Pfeffer out for a month, and the two McNellises being completely unapproachable, I found myself turning more and more to Chuck to find out what was happening with Xavira. He was always somewhere around her pen, either crouching down by the fence to watch her, or keeping an eye cocked as he worked with the elephants next door, or peeking out at her from one of the windows in the elephant house.

Chuck handled Xavira's sanitary requirements, and he had her bowel movements down to a science. "She takes her first one over here around eleven," he said one day pointing to the lower righthand corner of her pen from the walkway. He spoke out of the side of his mouth, swinging his head for emphasis.

"Then she takes another over there by the wall in the early afternoon, about one, one-thirty. I see her getting a little edgy, and rocking back and forth, and I say, 'Here comes the crap!' " Now that Xavira was getting ready to deliver—or so everyone hoped—it was his job to take a daily urine sample by applying a syringe to her puddle. He was also supposed to take a sample of the gooey discharge from her nipples, now swollen and rosy. That was a more delicate matter. He had to reach way under her broad belly and milk her like a dairy farmer, squirting the fluid into a plastic cup. Xavira had accommodated this with remarkable passivity until two weeks ago, when she recoiled from Chuck's touch and then charged him as though she wanted to boot him out of the exhibit with her horn. Chuck exited promptly. He was certain her orneriness meant she was ready to give birth. But he waited and waited, and nothing. All it meant was that Xavira wanted Chuck to leave his hands off her nipples. He still kept his eye on them, though, for any dribbles of milk that might suggest the baby was on its way.

For his part, Chuck's interest in Xavira was undiminished. As he talked with the visitors who were crowded around her pen, Chuck watched the rhino's every move as she trotted idly around her pen. It was as if he and not Billy were the father of her child. He was the only keeper in the entire zoo who volunteered for night duty during the rhino watch. The rest of the slots were filled by administrative and veterinary staff. Sometimes two or three days would go by before he left the zoo for a night's sleep. Chuck would drag himself around, barely able to keep his eyes open. The nightwork was unpaid, and he couldn't exactly say why he was doing it. "Well, hell, somebody's got to," he finally said.

If you peeled off his clothes, you could read Chuck's history on his flesh. He bore one scar that went nearly a foot up his belly, dating from age twenty-three, when half of his stomach had to be removed because of ulcers. "I take life too seriously," he said. "Everybody always says, 'Relax!' But I'm not much good at it." The rest of his wounds are attributable to his chosen career and to the fifteen years he had spent at the Philadelphia Zoo pursuing it. There is a puncture mark on his arm where a Chinese tree viper—a greenish, poisonous snake—got him some years ago. The snake hadn't been eating, and

Chuck had been trying to interest it in a "fuzz ball" or baby mouse. (Older mice on the menu are called, sequentially, "pinkies," "half-growns" and "full-sizes.") He was dangling the mouse by the tail from a pair of tweezers and protecting his hand with a metal shield when the viper sprang over the shield and sank its fangs into Chuck's arm. The police rushed him to the hospital. Doctors drew the venom out as best they could with a suction cup, and shot him full of antidote, antivenin. "I was more mad at myself than anything," he said. Some of the nerves in his hand were twitchy for a while, but he's better now.

Another time, a lusty otter jumped him and tore up his leg. "Nobody told me she was in heat," he groused. Otters barely weigh fifteen pounds, but their teeth are razor-sharp. Like most animals, they are fiercer when they are in estrus.

In recent years, however, he had devoted himself to the elephants. Unfortunately, they hadn't returned his affection.

"They tried to kill me twice," he said. "Or one of 'em did. Kutenga. She's crazy. One time she knocked me down and cracked my head open. She was down on her knees trying to get her tusks into me and finish me off when Jimmy and Danny backed her off me. The other time, she threw me down with her trunk and broke my ribs. She was about to sit on me when Jimmy came. He saved my life. Ever busted a rib? It hurts like a son of a bitch. Every time you breathe, it's like somebody stuck a knife in your lung."

"Why do you keep at it?" I asked.

"It's a challenge," he said. "It's definitely a challenge. When you win the respect of these animals, then you've really accomplished something. There's only one man who has learned to work with these animals when they were adults, and that's Dave Wood. I aim to be the second."

He paused for a couple of seconds, shaking his head. "See, the thing is, I really like elephants. And they like me, or at least all but one of 'em does."

Because he was a keeper in the elephant house, the Indian rhinos were also his responsibility. "Compared to them, Xavira is a piece of cake," he said. "There's not as much there. There's no telling what goes on in her mind. She's an airhead. But she grows on you. When she came in here, she was real spooky. You couldn't get near her. But now she's a good animal. She's manageable. She's lovable. Whenever you can get an animal to trust you, you've accomplished something.

"Jeez, if she'd only get on with his goddam birth. This has *really* been draining. With Gene out, I'm the only keeper on the rhino watch. I'm down with a cold. Eileen's been doing a lot of it, too. She's sick. If the baby doesn't come soon, we're gonna be dead. I never thought it would go this long. Hell, the pool runs out at the end of the month. If she don't drop by then, I got to give everybody's money back."

—— 13 ——

There's a nursery in back of the Rare Mammal House. It looks like it might be for baby humans, since it's one big kitchen, with a stove, refrigerator, and sink, and the room is painted in cheery colors. But in fact it's for raising other sorts of mammals—chimps, gorillas, aardvarks, you name it. The two lesser pandas, Jade and Rusty, had been there since they were born in June. Eileen's binturong, Ralph, had done some time here during the day when he wasn't curled up on her shoulder. Toward the end of October, the nursery added another distinguished visitor: the baby kangaroo Eddie.

One morning, Eddie had finally outgrown his L. L. Bean bag and graduated to a cardboard box. To ease the transition, the box was padded with his old imitation-sheepskin blanket. He was about knee-high now, and definitely a kangaroo. He had a long Pat Paulsen face, two powerful hind legs, two tiny front legs that seemed to have been transplanted from a hamster, and a thick tail. When he was older, he would be able to sit back on that tail and slash at his enemies with his powerful hind legs, but that was out of the question now.

His hopping was definitely coming along, but he still hobbled about a fair amount on all fours. In that position, with his nose down low over his little front paws, and his rear end raised up on his powerful back legs, he looked like a teenager's jacked-up Chevy. Unfortunately, on the slippery linoleum, Eddie had trouble getting out of first gear.

Still, he had mastered the cardboard box. He didn't slow down as he approached it, but charged at it at top speed like a pole vaulter, dived in head first, and then pulled his legs in after him. Once inside, he did a backflip to get his head up again.

Ann Hess was supervising him. As the zoo's surrogate mother, she was his daytime babysitter now that he had outgrown the Bean bag. "Hello, Eddie," she said in a pleasant singsong voice as she rubbed the sides of his face with her hands. "Hello, Eddie, hello."

A graying, fiftyish woman of great tenderness, Ann looked somewhat beaten down today. She'd been bitten on the index finger by a woodchuck last January, and the finger had stiffened up so badly she couldn't bend it. It stuck out painfully as she greeted Eddie. And this week, one of her two dogs had died: Blitz, a Doberman-like dog called a Rottweiler after the German town where the dogs were first bred. Blitz had had cancer, and Ann had been thinking seriously of putting him through chemotherapy, but finally realized that would be no use. Normally perky, she had been slowed by his death, and her voice was softer. In her grief, she had considered taking a special course in mourning for animals at the University of Pennsylvania vet center. She had finally decided to let it go with a few conversations with a psychologist. "It helps to talk about it," she said, "and to cry."

Blitz's death had been particularly hard to take because it was linked in a way to the death of her husband, Norm, who had been Bill Maloney's predecessor as head of animal services. It was Norm who had been honored on his retirement with the "Golden Turd" award for carrying "the heaviest load for forty years"—a trophy that still hung over Maloney's office door. The Hesses had gotten Blitz as a puppy eight years ago, just a month before Norm died of a heart attack.

Blitz would probably have been a favorite anyway. As ferocious as he may have appeared to others, he was always a gentle puppydog to Ann. Childless, she had come to regard Blitz as one of her offspring. He was like the oldest child who

helps with the younger ones. He had helped her raise a gorilla, an orangutan, drills, an aardvark, a chimpanzee, and a
leopard cub. "I always brought the zoo animals home in a carrying case first, so Blitz could sniff them. That way he could
get used to the new animals. It's like bringing home a newborn
baby. You have to let a dog know he is still part of the family."
Blitz played with the drills and the gorilla in her living room.
He once woke up to find the aardvark Aava napping on top
of him, and he took it very well. In return, Ann had granted
him special favors; she once took him to Disneyland.

Ann showered love on all her animal children. Of them
all, Kiki was probably her favorite. Kiki was a gorilla. She was
like a rambunctious little girl. Ann had pictures of her all around
her office, including an advertising poster for a car-seat man-
ufacturer that showed Kiki strapped into a car seat in Ann's
Pontiac, with the caption DON'T MONKEY AROUND WITH SAFETY.
Ann always put Kiki in a car seat when driving back and forth
from her house in the New Jersey suburbs. In many ways, she
had raised Kiki like a human baby. She had fed her formula
from a bottle, kept her in diapers until Kiki learned to pull off
the adhesive tabs, dressed her for bed in a blue sleepsuit with
feet, put her to sleep in a playpen—with a Plexiglas lid so Kiki
couldn't climb out—and for exercise, she strapped her into a
Jolly Jumper, a trapezelike device that allowed Kiki to bounce
in the doorway. She said that a baby gorilla required a little
more attention than a human baby. "You can leave a kid for a
few seconds, but not a gorilla. If you turn your back on a gorilla for a second, you lose your drapes." Kiki also craved affection. "She would put her hands out for me to come pick
her up, just like a kid," Ann remembered. "Or she'd grab
on to my leg and hug on to me. And she giggled when I
tickled her."

In the end, Kiki was sent away to the Stone Zoo in Stone-
ham, Massachusetts. Ann had been up there a few times to
visit her. Kiki still recognized Ann, although as time went by,
it always took a little longer to place her.

Eddie curled up for a nap in his cardboard box, and the
two lesser pandas took the stage. They looked more like a hy-
brid of a fox and a raccoon than like their heralded cousins

the giant pandas. They had foxy faces and ringed raccoon tails. Their backs were reddish brown, and their bellies charcoal black as though they had been singed. The pandas played together like a pair of oversized kittens, rolling all over each other madly until I happened to flip a page of my notebook. Then they both stopped as though a gun had gone off, and came over to investigate. "If they hadn't seen humans before," Ann said, "they'd bite your hands off." As it was, they contented themselves with undoing my shoelaces and rubbing their behinds against my pantleg. "They're territorial markers," said Ann. "They like to squoosh around and urinate on new things." I backed off, but the pandas held their fire. "If one pees on something, then the other goes on top of it," Ann added. "It's a dominance thing." I gave each of them a pat on the head and a blue hard-rubber pretzel.

Mike Barrie came in, trailed by the A.P. photographer Amy Sancetta and PR director Debbie Derrickson. They wanted to take some shots of the veterinarian and his foster child. Eddie was overjoyed to see Mike. His feet clattering on the linoleum, he bounded over to Mike and hopped up onto his belly, and banged his head against Mike's belt buckle. "Pouch! Pouch!" Mike yelled, laughing and protecting himself with his hands. "I think he's imprinted on me," said Mike bashfully. Eddie was hopping around the room manfully. "See? He hops like a champ." Mike offered him his little finger to suck on to quiet him. Eddie leaned back on his tail, forming a tripod effect with his two hind legs, and took it earnestly. Then Mike fetched a bottle of formula from the refrigerator, warmed it up under the hot-water tap, and offered it to Eddie, who was very interested. Cradled in Mike's arms, Eddie pulled away on the bottle while the photographer took photo after photo. The picture of Mike and Eddie would go out on the A.P. wire the next morning. "We are the happy-news suppliers for the Delaware Valley," said Debbie Derrickson. When Eddie was finished, he made a dash for his box once more and plunged in. In his excitement he forgot the final backflip, and ended up head down in the box with his two big legs sticking up in the air. Mike had to straighten him out while we all laughed. "Yessir, that's my baby," said Mike.

—14—

I had thought I would stay around the zoo till the rhino baby came, but after two weeks of waiting, I began to reconsider. Every day I would scrutinize Xavira, sometimes for hours at a time, looking for some hint of her plans for delivery. Chuck Ripka had told me to pay particular attention to the slot for her tail in her rear armor. If that began to widen, the baby was on its way. My eyes were fairly glued to Xavira's rear end, but it never showed any change. The slot remained as snug as ever. In fact, Xavira never changed at all. In my eyes, she remained one of those perpetually cheery maidens, ever smiling, ever bouncy, even in the face of the most dire calamity. Day after day, she would wander contentedly around her pen. If she had been human, I expect she would have hummed.

Every once in a while, she would swing her head around awkwardly to look over at Billy, although the two never made eye contact as far as I could see. More often, she would rest her head on the wall by the elephant yard to gaze at the pachyderms. Sometimes the elephants would come by and sniff her all over with their trunks, like doctors probing with stethoscopes. I imagined they were just as curious about the delivery as I was.

Finally, my wife called from Boston to ask when I was

coming home. She had been awfully good about my pro-
tracted visits to the Philadelphia Zoo, but there was an edge to
her voice this time. She said that our two-year-old was sick,
and that she was coming down with something, too.

I went out to take another look at Xavira. She was jogging
around happily. Then I checked once more with Wilbur Amand,
who said that there was really no way of knowing when the
rhino was going to drop. There was so little experience of rhino
births in zoos that they couldn't tell what the time limits were.
It could easily be a couple more weeks, he said. At hearing
that, I rolled my eyes in despair, called my wife, and told her
that I would catch the next flight back.

I checked in with the zoo every day while I was gone. Three
days went by, and nothing had developed. Finally, I called on
Sunday, November 5, and Debbie Derrickson told me she'd
heard that Xavira was acting jumpy. I asked her to check into
it and call me right back. She called back five minutes later.
"You'd better get down here," she said. I caught the next plane
down, took a taxi from the airport, and pulled in to the zoo at
ten o'clock at night. There was a security man waiting in a golf
cart to whisk me to the elephant house. I asked breathlessly if
the baby had come yet.

"Sure has," he said.

I groaned. "What is it?" I asked.

"It's a rhino."

"Yes, but a boy or a girl?"

"Now, that I couldn't tell you."

We didn't speak the rest of the way. I had never been in
the zoo at night before. It was surprisingly quiet, and a little
spooky. A full moon was out, and it cast soft shadows across
the garden. Occasionally, I heard a muffled roar, or a quiet
lowing, or a snort by the animals as they bedded down for the
night indoors. Mostly, I was aware of the trees creaking in
the wind, and the cold air on my cheeks as we rushed along
the sidewalk.

He took me around to the back of the elephant house,
and I hurried in. It was bright inside, and there were about a
dozen people standing around Xavira's cage. Ann Hess, Chuck
Ripka, and Dave Wood were squeezing in the doorway to the
exhibit area, trying to get a good look. Eileen Gallagher stood

behind them blinking away some tears. Xavira was lying on her side, clearly spent from all the exertion. And there was a little bundle by her side nosing at her teats. It was the rhino baby. It was alive.

Mike Barrie was leaning against the wall, watching. He smiled when he saw me. "Hey," he said. "You missed it!"

"I know," I said glumly. "What happened?"

"Well, he just kinda flopped out."

"He?"

"Yeah, it's a boy."

"And you should see the equipment on him," added Dave Wood. "We're calling him B.J. for Billy Junior. He takes after his dad. He's only two hours old and he's already had his first erection. He's a breeder, that boy."

Mike filled me in on the basics. At about three that afternoon, Xavira had undergone a marked change of character and chased everybody away. They managed to get her inside to her indoor pen, where she tossed her head around wildly and couldn't seem to get comfortable. Her waters broke at a little before six, flooding the cage. The labor got intense around seven, and at seven-forty, the baby's front legs and nose peeked out of Xavira's vagina. Fifteen minutes later, B.J. managed to work his head and shoulder out. Finally, at 8:05 he hit the ground. Xavira whipped around as if to see what was happening back there, and burst open the amniotic sack. The only problem was that Xavira had delivered the baby on a heap of dung—"That's Murphy's Law," said Mike, "always the worst possible place"—so the vets had to pull him out and clean him off, then give him back to her. But she was so calm about the whole thing, she didn't mind. This also gave the vets a good opportunity to jostle B.J. a bit—an important element in helping a newborn animal to adjust to the world. "You want to get their senses going," Mike said. "All we're waiting for now is for B.J. to get some of that milk." The baby rhino was quite eager. As we watched, he took a couple of swipes at the nipple, but he couldn't quite get hold of it. "At least he's in range," said Mike.

Just then, B.J. found his target, the big red teat disappeared into his mouth, and there was a contented, sucking sound like a bilge pump.

"Hey, he's latched on pretty good," said Mike.

Wilbur Amand greeted me and led me around to the inner hallway so that I could get a better look at the newborn. The atmosphere was very much three A.M. on New Year's morning. Eileen had strapped on a plastic hippo snout from Halloween. "They didn't have any rhinos," she explained. The big moment had come and gone, but everybody was still far too excited to leave. Their attention was fixed on the baby rhino, still sucking away at his mother's teats.

B.J.'s whole body was about the size of his mother's head. Like a newborn human baby, he was oddly proportioned. His own head, eerily fishlike, was big for his body, which in turn was big for his spindly limbs. I couldn't imagine how he would ever walk. But he managed it in all of thirty minutes. Speed is critical in the wild, for newborns need to get moving to escape predators. It was for the same reason that Xavira gave birth in the evening: It allowed her offspring to become adjusted to life outside the womb under cover of darkness.

I later saw the tape of B.J.'s struggle to lift himself off the ground. He looked like someone trying to get up on skates. He had the general idea, but was utterly unsure of his technique. After much struggle, he managed to prop himself up on his front limbs, but he couldn't get the back ones in place. Then he would get the back ones under him, but the front ones would slip out. Finally, he got the whole job together, and he stood, amazed at the view from that elevation.

After we had gaped at B.J. for a while, Wilbur led everyone up to the conference room at Penrose, where we continued to watch the spectacle on the video monitor. "Look at them ears," said Dave Wood as B.J. stood up to reveal his little tufted donkey ears in profile.

"They're great ears," agreed Eileen with a smile. "But I like his heinie."

"I say the whole thing's not bad," said Wilbur Amand, "not bad at all."

The monitor showed B.J. resting his chin in the hay.

Chuck Ripka took a seat next to me at the conference table. He sat there silently with a big grin on his face, nodding his head. "It's hard to believe this moment," he said. "I'm amazed, just amazed. My heart's still pounding. After all this waiting, we have a baby rhino. I can't believe it!" He was proud to say that he was the one to come up with the name B.J. He had been thinking about it for months. "I wanted something short," he explained, "and I thought he should be named after his father. He looks like Billy, don't he? He eats like him anyway."

Dave started the videotape going to look at the birth one more time. Xavira's rear was to the camera and a little out of focus. She seemed frisky. She would stand up, then sit down, then stand up again. Finally, a little white bulge appeared in her vagina, like the beginnings of a chewing-gum bubble. It puffed out, then pulled back, then puffed out again; and finally bulged with the distinct image of a single pale hoof. As we watched, Mike started humming the opening bars of *Thus Spake Zarathustra.*

"I was a cheerleader in high school," Eileen yelled over to me, "and I was up on the table screaming, 'Push! Push! P-U-S-H! Push!' "

"You were screaming your head off, Eileen," said Dave.

"Okay, so I got a little excited," said Eileen.

The events proceeded rapidly—four hooves appeared, then a head and a shoulder, and then the whole baby rhino dove out and landed in a heap on the hay-strewn floor.

"We have rhinoceros," said Mike in a Mission Control voice.

"When the baby came, I called my parents," Eileen said. "And I just yelled at my father, 'It's a boy! It's a boy!' "

Keith saluted Xavira's image on the TV. "Thanks for all the fun, girl."

"And it's your birthday, too," shouted Dave.

It was Keith's thirtieth birthday today. He had been out shopping for a birthday present for himself when he was beeped and told to come into the zoo because labor had begun.

"What a great present!" shouted Eileen. "I'd kill for a baby rhino birthday present."

With that, Dave Wood passed out some paper cups and brought out the bottle of Great Western Extra Dry that he had been saving for nearly two years. It was time for a toast. He poured champagne all around. The place went quiet. Dave gestured toward the screen, now filled with young B.J. sampling a little hay. "We're looking at history here," he exclaimed. "In the one-hundred-eleven-year history of the Philadelphia Zoo, this is a first. It took a long time but, damn, it was worth it."

We all raised our cups.

"To B.J.!" Dave exclaimed.

"To B.J.!" we repeated in chorus. And, our eyes on the TV screen, we all drank our champagne.

Winter

❄

——— I ———

Wolf Woods is the zoo's Siberia. It's a remote spot at the far end of the zoo from the main gate, past the farmlike Children's Zoo, and along the fence by the train tracks that girdle the zoo. One might think this is where the least popular animals are exiled. The wolves are quartered here next to the Cape hunting dogs—fierce spotted mongrel-like creatures—and some miniature deer called muntjacs.

Although the wolf has become somewhat rehabilitated in the popular mind in recent years, it has had a lot of ground to regain. In the Middle Ages, wolves were considered incarnations of the devil himself, preying not only on sheep but on women and children, as in the tale of Little Red Riding Hood. In the United States, the wolf has historically symbolized the brutality of the uncultivated wilderness. At the turn of the century, the Bronx Zoo's William Temple Hornaday called wolves the most "despicable" creatures in North America. "There is no depth of meanness, treachery or cruelty to which they do not cheerfully descend," he declared. Across the country, hunters were encouraged to shoot all wolves on sight.

More recently, through the efforts of such naturalists as Adolph Murie and David Mech, wolves have come to be seen as more noble creatures that form monogamous bonds that

247

look not a little like human love matches, establish elaborate social hierarchies, and actually strengthen the herds of elk and caribou they prey upon by culling the sick and infirm.

The "woods" at the Philadelphia are a narrow strip, about twenty yards deep and forty yards across, that curves around between the walkway and the picket fence. The area is thinly planted with a few bushy pines and strewn with boulders. The grass has long since been trampled into dust by the wolves as they raced back and forth.

During the warmer months, the wolves are usually sacked out, panting in the heat, about ten yards apart from each other in accordance with some internal social code: the male, Taboo, taking the higher ground under a tree by the far wall, and his three sisters, Trillion, H.B., and Koshie, scattered like vassals around him. They are all handsome animals. Wolves are about the size of German shepherds but, as Barry Lopez pointed out in *Of Wolves and Men,* we make them bigger in our imaginations. Their coats are thick enough to turn into knitting wool, and grow shaggy in winter. Among the Philadelphia Zoo's pack, the colors range from the arctic white of Koshie to the cinnamon and gray of H.B. All the wolves' coats run darker along their shoulders and spine. Compared with dogs, their jaws are longer and more powerful, and their bodies more streamlined. But the real difference lies in this: Wolves, even in captivity, take possession of their environment, whereas dogs merely traverse it.

While most of the other animals at the zoo have to be tended, if only to be shifted in and out of doors, the wolves are pretty much left alone. They stay out, rain or snow, night or day. A keeper tosses them some meat and bones once a day, and that's it. Unlike the elephants, the captive wolves don't need to integrate any humans into their dominance hierarchy, which is even more finely detailed than the elephants'. For humans are almost never present, except for the droves of schoolchildren that cluster about the exhibit and howl. I might not have been drawn to them myself except that the vets happened to bring me along one day in the late fall when they made their regular visit to the wolves to "socialize" them, as they put it.

Eileen first mentioned the practice one day in her lab. She

said that the wolves had all come from a Purdue University ethologist named Dr. Erich Klinghammer in Battle Creek, Indiana. Klinghammer had recommended that anyone needing to interact with the wolves should do some howling with them every once in a while.

"Howling?" I said.

"Yeah," said Eileen. "It helps the wolves get used to you."

"Do you do that?"

"Sometimes," she said. "I sound like a sick duck, but Dave Wood and the vets are a little better. You're also supposed to get your hands on them, give them a rubdown or something. I find that part a little scary, but Keith and Mike and Dave do it. I guess it helps, who knows?"

At the beginning of November, while the zoo was still waiting for the rhino birth, it had come time for Dave and the vets to pay a call on the wolves once more. They drove out to Wolf Woods in their pickups, pulled on overalls to protect their zoo khakis from muddy pawprints (and to protect the wolves from protruding belt buckles), then they passed through some hollow Gunite boulders and headed into the exhibit.

As they went, I noticed a short, heavily clothed woman over to my right. She was sitting on a stone bench that had been almost completely enshrouded by an overgrown yew bush. Her eyes scarcely left the wolves the whole time I watched her. She looked like a boulder wrapped in wool. She had on a blue ski parka, heavy work boots, a wool ski cap pulled down low over her eyes, and a thick scarf wound around her neck and up to her nose, leaving just a narrow slit for her eyes. Seeing Dave and the vets climb into the exhibit, she stepped up to the fence to watch. I realized I was in the presence of the Wolf Lady.

Her name was Janet Lidle. I had heard about her from various keepers and administrators already, but I had not known what to make of the stories. Most of the zoo people thought she was a little peculiar. To judge by what little I could see of her under that mound of clothing, I would guess she was in her forties, and she was short and thick-set. She had the obdurate look of an Eskimo squaw.

I strolled over to say hello. It was a little difficult to communicate with her under all that clothing; the sound was heavily muffled in both directions. Plus, I'd been told she was hard of hearing, and blind in one eye. These disabilities, I learned later, had in a sense brought her to the zoo. Janet had worked for several years as a clerk at the post office, but she was pensioned out early because of her infirmities. One day around ten years ago, she had begun writing a short story about some Alaskan settlers who are set upon by a pack of wolves. She came out to the zoo to do a little research. The research became extended. For the next five years, she returned to the zoo once a week. Then she started coming every day except weekends, when she was put off by the crowds. She still followed the same schedule almost religiously: Arriving at eight-thirty to take up her position here on the stone bench, she would stay until three or four. She still hadn't finished that short story. She said that she wasn't the only one to have a "fascination," as she put it, for an animal. There used to be a man named John who was equally devoted to the Siberian tigers, but he didn't come around anymore. "We were the Cat Man and the Wolf Lady," she said and gave a strange, cackling laugh. "People don't expect that. Usually, it's the Cat Lady and Wolf Man."

Now, as Dave Wood and the vets entered the enclosure, I asked Janet if she wanted to go out to tussle with the wolves herself.

"Not me," she said with another cackle. "Number one, I wear glasses. Two, I don't have workman's comp, and three, I don't want the wolves to notice me." She didn't want to influence their behavior.

Janet and I looked across the dry moat and into the exhibit. Dave and the vets looked incongruous in there, as staffers always do when they break the plane of illusion and mingle with the animals. The three men started to howl, tentatively at first, then lustily. Sounding more like the drunken caterwauling of college kids, it was a pathetic imitation of the great, full-throated songs of the wolves. The three wolf sisters came over, probably to see who was making such a racket, and gave the three men a sniff. Their brother, Taboo, the hulking male, hung back. Keith had said that when playing with the animals, it was important to get the wolves down on their backs quickly. Wolves value height and accord it, as humans do, a mark of status. When a higher-ranking wolf approaches a lower-ranking one, the former will have his tail out straight and his ears up; the lower-ranked wolf, unless he wants a fight, will immediately make obeisance by dropping his tale, lowering and pushing back his ears, slumping his shoulders, and generally acting like a puppy about to get thwacked with a rolled-up newspaper. So now the three men immediately sank their hands deep into the thick coats of the wolves and rolled the animals over their backs. The three sisters appeared to be in ecstasy as they writhed on their backs, jaws wide open and emitting playful yips and growls interspersed with happy gargling sounds. The men stroked the wolves' bellies and yowled.

Suddenly, the jolly mood was shattered. One of the wolves jumped to her feet, and set upon one of her sisters before she could leap up to defend herself. "That's Trillion," said Janet. "She's been waiting for months to get back at H.B."

Trillion was the smallest of the three sisters, lean and scrawny. H.B., by contrast, was most impressive—nicely filled out, with a handsome coat of gray and cinnamon brown. Janet explained that H.B. had been at the top of the hierarchy for the last few months, and had been ragging the lowly Trillion. At the slightest insubordination, H.B. would fly upon her in a rage. Now, however, the intrusion of the three men had altered the power balance. At seeing H.B. lower herself before the men, Trillion saw her opportunity, and she was making the most of it. She bared her fangs and went after her sister. The third sister, Koshie, saw the sudden change in the rankings and immediately came in on Trillion's side.

There was a tumult of dust and flying fur in the exhibit as Trillion and Koshie tore into their sister. The three wolves thrashed about at the feet of the men for a minute or two. The men pulled back. They wanted to stay out of this. Seeing that she was outnumbered, H.B. tried to make a run for it. Trillion and Koshie dashed after her, caught up with her quickly, and hacked at her flanks with their fangs. For all the frenzy of the battle, the three wolves were completely silent. All we could hear was the pattering of their paws as they raced across the exhibit. That made Janet edgy. At first, she had treated the scuffle as a diversion. Now this looked like war.

"Throw a rock at them!" she yelled at the men.

Dave grabbed a rock and flung it at the wall of the dry moat below us. The rock pounded hard against the cement and broke into bits. The wolves paid no attention. They were hidden from our view at the base of the moat. We could hear them rolling and tumbling as before.

"This is rather serious, gentlemen!" Janet yelled. She was trying to stay calm.

Dave flapped his arms and yelled, "Hey! Hey!" sharply at the wolves, but they paid no attention. The vets looked on with concern.

"Cut it out!" shouted Keith. He threw a rock at them awkwardly. It missed by a wide margin and bounced harmlessly across the ground.

We heard nothing except the sound of scraping paws. "If they're totally silent," said Janet, "it means it's really serious."

"I think we'll leave," said Mike. He knew there was nothing the vets could do except get hurt. Even if they could pull the animals apart, as soon as they were brought back together again, they would simply take up where they had left off.

"I think H.B.'s bleeding," said Janet. The wolves had reappeared on the crest of the exhibit above the moat. Some blood was trickling down from a corner of H.B.'s ear. "Trillion's tried to rip her ear off!"

"Well, the stones aren't doing anything," said Dave. "And I'll be damned if I'm going to get in there and wrestle with them."

"No, no," said Janet, "You don't break these things up personally."

Dave and the vets retreated from the exhibit, casting nervous backward glances as they went. Trillion continued to race H.B. back and forth, nipping at her hind legs. Whenever she fell back to rest, Koshie took over. Finally, H.B. slowed down. Before the fight, she had walked proudly erect. Now she was hunched over, her tail between her legs, as if to protect herself from further blows that she was too tired, and too beaten, to fight off.

"Looks like she's being submissive," I said.

"I would be, too," said Janet, "if I had a couple of wolves after me."

Dave Wood had come to join us at the fence. "We had ourselves a little dominance struggle," he said. "That's why we left. We didn't have anything to do with it. We didn't exist for them."

"Wolves are like people," said Janet. "They see who is on the bottom and then they jump on them."

Dave and the vets got back into their pickups to ride back to Penrose. As they left, Trillion was marching back and forth at the center of the exhibit like a champion prizefighter. H.B. had slunk off to a far corner like a dethroned and embarrassed despot. After her regal splendor, she looked utterly defeated, mere riffraff. It hurt to see her brought so low. Koshie, true to her standing, occupied the middle ground between her two sisters, alternately glaring at H.B. and looking back anxiously and deferentially at Trillion. And above them, virtually un- moved since the whole episode began, was Taboo, the solitary male, whose affections the three sisters were fighting for. This February the females would be coming into heat for the first time, and Taboo would attain sexual maturity. It would be time for selecting a mate. It didn't matter to them that he was their brother. Now he looked on dispassionately, apparently amused by all the ruckus.

When I returned that afternoon, Janet was asleep on her bench and the wolves napping. Taboo had taken up his cus- tomary place by a rock deep in the exhibit, and the three sis- ters were slumbering their usual ten paces away from him. Janet woke up as I approached. It took her a moment to place me. "The wolves usually sleep in the early afternoon," she ex- plained. "So I get some rest then, too."

We looked out at the wolves, who still maintained their social distance from each other in sleep. "It's rare to find them sleeping close together. Not like *them*." She gestured derisively toward the Cape hunting dogs. "They sleep together in one big ball."

We talked about the fight. "It was pretty fierce," Janet said, "but I've seen plenty worse. A couple of wolves had to be taken out in 1977 because of dominance fights. That was Suncleep and Fair Lady. They nearly killed each other. They shipped them off to a breeder in upstate New York. And a couple of females, Fire Eye and Sunbeam, went at it pretty bad in 1981.

In 1984 there was a nasty fight when these four"—she looked at the current residents of Wolf Woods—"came in. That really threatened the hierarchy."

Janet explained the pack genealogy. It was almost as complicated as the British royal succession. The short of it was that Taboo and his three sisters hadn't been whelped by any of the wolves Janet had mentioned. Because of a ten-year age gap, they were, amazingly, four generations removed from Shy Boy and his mate's sister, Fire Eye, and two generations from Sunbeam, their great-aunt. That group had died out earlier this summer. The current wolves' familial line had developed from one of Shy Boy's progeny at Dr. Klinghammer's breeding ground in Indiana. The call had gone out to Dr. Klinghammer for some pups in 1984 when it became clear that the incumbent group was starting to show its age. Fire Eye and Shy Boy had some arthritis, and Sunbeam a tumor. It was largely a coincidence that the two groups of wolves were related at all.

When Dr. Klinghammer drove up that summer with his litter of four three-month-old wolf pups in a large wire crate, no one quite knew how the youngsters were going to be received. Janet had been consulted by Dietrich Schaaf, the curator of mammals, and she had voiced some uncertainty about the undertaking. Clearly, the puppies would upset the established order, but how would it be resolved? At the time, Fire Eye and Shy Boy formed the alpha pair, but Fire Eye, increasingly hampered by her arthritis, had been struggling hard to put down the regular insurrections staged by Sunbeam. Fire Eye was snow white, compared with Sunbeam's dull gray. Her battles with Sunbeam had put a kink in her tail, scarred her muzzle, and taken a chunk out of one ear. Sunbeam was younger, closer to the pups in age and by blood. Would she take them over, possibly in hopes of developing a bond that might assist her in her assaults on Fire Eye? Would Fire Eye regard the pups as a threat and exert her power over them, maybe killing them? Would both bitches spurn the youngsters?

The answer came quickly, and it was most unexpected. In all her years of watching the Wolf Woods, it proved to be the most memorable event Janet had ever witnessed.

Far from being threatened by the newcomers, the grizzled

Fire Eye welcomed them as her own. Janet described that first look in one of her *Pages from a Wolf Log,* a regular feature of a quarterly newsletter called *WOLF!* that she puts out for a few hundred subscribers who share her enthusiasm for wolves: "As soon as she saw the pups in the pen, Fire Eye's expression changed from fearful aggression to wonder and delight. Wide eyes fixed on H.B., she ignored the people [Dr. Klinghammer and the zoo staffers had gathered to watch the encounter] and headed straight for the nearest pup. H.B. backed up until a fence stopped her. The big black nose touched the little black nose. H.B.'s apprehension disappeared. . . ."

The other pups took their cue from H.B. and looked to Fire Eye for comfort. Fire Eye nuzzled them tenderly. Sunbeam kept her distance. When two of the pups were playing with a piece of meat, Sunbeam pounced on one of them and nipped her. Fire Eye dashed to the rescue and chased Sunbeam away. Later, Sunbeam attacked Taboo. Once again, Fire Eye intervened. As she looked on, Janet sensed that Sunbeam viewed the pups as she did the pigeons and squirrels that strayed into her pen. She stalked them like prey. Only Fire Eye's vigilance saved the pups' lives.

Fire Eye was ornery toward people. She had taken a hunk out of the rear ends of two of her keepers, and had always acted hungry for more. "She'd have bitten Saint Francis," said Janet. But now, whenever the zoo staff came into her pen, she didn't try to bite any of them. She was concentrating too hard on protecting the young from Sunbeam. The zoo authorities realized that Sunbeam had to go, and Dr. Klinghammer agreed to take her back to Indiana with him. When Fire Eye and the pups were introduced once more after that first testing of the waters, the old wolf took to them instantly. Shy Boy, true to his name, hung back, but the young Koshie, not knowing any better, sought him out, pawed his nose, and gnawed on his muzzle to beg for food in accordance with an ancient ritual. Responding, he produced her dinner by regurgitating. Koshie and Trillion rolled around in the mess for a few minutes and then nibbled at it. In the coming months Trillion would establish a ritual of giving Shy Boy a yank on the tail, after which he would immediately deliver her dinner. "It was like she was ringing the dinner bell," said Janet. For her part, Fire Eye took it upon herself to protect the pups from all intruders. Once, when Dave Wood's late predecessor, Rick Beyer, approached the pen, she growled fiercely at him and then snapped her jaws shut so hard that the sound echoed around the zoo. Beyer kept his distance. He got the message: These pups were hers.

"How do you explain it?" I asked Janet.

"Sometimes I think we try too hard to explain things," she said with a shrug. "Fire Eye had a strong maternal instinct. That's all you can say."

At first, Fire Eye and Shy Boy seemed rejuvenated by the puppies. There was new spring in Fire Eye's step, and Shy Boy was more outgoing than Janet could ever remember. "He romped with [the pups] as they grew," Janet wrote, "and even displayed again his leaping, whirling, head-tossing, tail-twirling 'dance.' "

But a year passed, and then another, and the pups stopped displaying their gratitude. Trillion and H.B., the two whose dominance fight I had just witnessed, started to turn on their benefactor. They would nip at her, sometimes slashing deep into her side. Fire Eye was remarkably patient. Even though Trillion was abusing her, when Trillion became frightened at the sight of a telephone company repairman working on the line outside the zoo, Fire Eye ran to comfort her. But the scuffles continued, and before long Fire Eye was acting like the omega wolf.

One night in January, one of the wolves, almost certainly Trillion or H.B., wounded their adopted mother seriously on the thigh. She recovered slowly. The pups continued to bedevil her as she healed. In the spring she spent most of her time sleeping. In late May one of the pups injured her again, and this time she had to be removed from the group. Now she was no longer able to defend herself, and for her own good, Keith Hinshaw put her to sleep. By now, Shy Boy had developed a bad case of arthritis. He could still hold his own with the pups, but it was only a matter of time before Taboo, who was growing more adult every day, would challenge him for the alpha spot. Seeing how decrepit he had become, Dietrich Schaaf decided to put him down in July. Janet wrote obituaries for those wolves in her newsletter.

"She was easier to love than to like," Janet wrote of

Fire Eye. "As straight-forward as a steamroller and just about as subtle. . . . She did have a beautiful gait, effortless and free-flowing. She displayed it most often in her 'Where is that keeper?' pacing routine. I imagined her singing softly to herself as she padded back and forth, 'All I do/ Is dream of food/ The whole day through.' Indeed, the great loves of most of her life were meat, bones, and Shy Boy—in that order. How food oriented was she? Twice I saw her drag Shy Boy backwards toward the meat when feeding time arrived while they were tied in mating."

As for Shy Boy, Janet wrote, "If meekness is the sole qualification for inheriting the earth, Shy Boy has a better claim than most people. He was the closest thing to a saint I ever knew, except for his bushy tail. . . . Great as was his physical beauty, lovelier by far was his personality. He possessed in abundance that quality G. K. Chesterton ascribed to the Celtic gods: an 'unbearable beauty that broke the hearts of men.' "

—— 2 ——

Eddie, the baby kangaroo, died one Saturday in December. Arlene Kut at the PR office told me the news when I came in the following Monday morning.

Arlene had taken over from Debbie Derrickson, who had gone on to a new job with a large PR firm in downtown Philadelphia. Arlene was one of the jolliest people you could ever meet. She rarely went two sentences without giving a wonderful big-hearted laugh, but this had been a hard week, and even she seemed drawn. The zoo had lost Eddie, a springbok, a zebra, and a beautiful black jaguar all in the same week. As soon as I heard about Eddie, I hurried over to Penrose to see Eileen.

She was at her microscope, as usual, when I came into the lab. And she wore an exotic seed necklace that John Groves, the curator of reptiles, had brought back from the Amazon, where he had gone on a collecting trip. Eileen had taken care of his parakeet and his fish while he was away.

She seemed okay. She didn't even mention Eddie at first, but talked instead about young B. J., the rhino baby, over in the elephant house. "He's really doing good," she said. "He's so cute." But as she talked, I could tell that something was bothering her. So I asked her about Eddie.

259

SO. ST. PAUL PUBLIC LIBRARY
106 3RD AVE. N.
SO. ST. PAUL, MN 55075

"He died in the hospital," said Eileen, referring to the Penrose Annex. "He'd been sick for two weeks. He had terrible diarrhea, and he'd stopped eating. He hadn't been eating well for a while before that. It was so bizarre! He'd be sick as anything one day. Then the next day he'd be up hopping around. He'd be looking around, cheerful. It would go like this, one day bad, the next day good, then the next day bad again. I remember on one of his good days, somebody left the door open, and Eddie hopped right through and ran down the hall. Mike saw him go, and he couldn't believe it. The next day, Eddie couldn't stand up.

"Friday was a good day. He was fine. Then on Saturday I came in to work on something, and Ann Hess ran in and told me that Eddie had stopped breathing, and his heart had stopped, too. We raced in there and worked on him for forty-five minutes. But we couldn't bring him back. He was gone."

I asked how she was feeling.

"Not too bad, I suppose," she said. "The thing is, we had been expecting him to die for two weeks. I'd been crying my eyes out for him every day. But now that he's gone, I have no more tears left. I guess I'm relieved for him. He's not going to suffer anymore. But his death leaves a big hole in my life. Seems like all I had was Eddie and my dog."

"But you've got friends," I said, trying to cheer her up.

"Oh I know. But I'll really miss Eddie. I really loved the way he'd come hopping up to me, all happy, and grab on to my leg and give me a hug. He was a really great kangaroo."

"How about Mike," I asked. "Is he okay?"

"He wasn't here when it happened," she said. "It was his day off. But he came in the next day. He was joking around, but I knew he was heartbroken. We all loved Eddie. But Mike had spent the most time with him. He had him just about every day for the last two months."

"Any idea what went wrong?"

"Dr. Snyder did a necropsy. The intestines were shiny on the inside, like they get with chronic diarrhea, and the kidneys were a little pale. But we're waiting for the slides. But you know, this almost always happens with kangaroos. The mother rejects them from the pouch and it looks like nothing's wrong with them. But they always die. It's like the mothers know something that we don't. We always think that we can pull them through, but we never can."

I heard the vets talking in the hallway and heading into their office, so I ducked out of Eileen's lab and went around to see them. Keith was looking perfectly calm as usual. "Well, you're here just in time for the kangaroo funeral," he said. Mike was behind him, looking drawn but not unhappy.

"I thought you didn't *have* funerals," I said.

"Well, we don't. But I thought that for one brief second we might have one for Eddie. It was the same with Massa."

"Sounds like you two are bearing up okay," I said.

Finally, Mike spoke. "Well, Eddie had been sick for two weeks, so we had some time to prepare ourselves. Plus, being a vet, I've seen so many animals die, I've kind of gotten used to the idea."

"But Eddie must be different."

For the first time, Mike looked sad. It was as though he hadn't been thinking about Eddie before.

"Yeah," he said. The words didn't come for a second or two. "I was very disappointed. He'd made it further than any other kangaroo we'd tried."

"Do you know what went wrong?"

"Well, we had to put him on antibiotics for the diarrhea. But they had a bad effect on his kidneys. He died of septicemia. That's when the kidneys aren't filtering out the bacteria, so it runs through your bloodstream and poisons you. So we had to take him off the antibiotics, but that left him completely debilitated from the diarrhea."

Keith picked up the story. "He'd have a good day, then a bad day, and the reason was that we'd put him on an IV to correct the pH and the protein levels of his blood. But once he got on the IV for a day, he felt so good he'd start jumping all around the place, so we couldn't keep him on the IV anymore, and we put him on oral fluids. Then he'd go downhill again."

"It was terrible to see him keep having the bad days after the good ones," added Mike. "That was so discouraging. My wife Kathy's really sad about it, too. Since she's a veterinarian,

262 ──────── The Peaceable Kingdom ────────

she loses a lot of patients, too, but this still hurts . . ."

His voice trailed off. The silence hung in the room for a while. He looked up at me bleakly. "I don't know what to say. I don't know how I'm feeling. Eddie's only been gone since Saturday."

—— 3 ——

Winter had taken hold of the zoo by now. The earth was frozen as hard as concrete, and the trees looked brittle. The few animals that were still out—cheetahs, barasingha deer, and others—huddled in the lee of their sheds, trying to stay warm.

But no matter how low the temperature dropped, it was always warm and steamy in the Pachyderm House. It was like a greenhouse with cement walls. Water leaked from the radiators and dribbled across the floor. Down at one end, the hippos were splashing about their pools, noisy as ten-year-olds on a summer's day. The tapirs in the next cage squeaked and barked. And up at the other end, the elephants lumbered about noisily.

But Xavira's cage was quiet. With all the straw strewn across the floor, and everything so peaceful, it looked like a Nativity scene. B.J. was asleep with his back up against the wall. In that position he reminded me of my own daughter, curled up against the bumpers of her crib. Xavira stood protectively over him, munching on hay like a cow. Billy roamed about four stalls down, occasionally peering over the wall at his progeny. As B.J. stirred in his sleep, his donkey ears twitched occasionally, and his eyes opened wide for a moment—as though to see if

263

the world was still the way he remembered—and then snapped shut again.

He looked a little bigger than he had at birth, which was only natural, since he was drinking six gallons of Xavira's milk a day. In fact, he had gained a hundred pounds. But he was still so much smaller than his mother that it was hard to tell for sure if he had grown. His hooves had lost their velvety whiteness and toughened up noticeably, but his leathery folds were still the ruddy brown of a newborn's. Dave Wood told me that B.J. had chased him out of the exhibit a few days ago. "He gave me a choice: I could leave, or he could take me off at the kneecaps." Dave left. "He's little, but he's built like a rock," he said.

Gene Pfeffer stood by the wall looking at the new rhino, and I asked him how he was coming with the elephants.

He took a long drag on his cigarette. "I was off for a few weeks in October, and when I got back in November I was back at ground zero," he said gloomily. "They were out there sniffing me like they didn't know who I was. It was like—'Who the hell is this guy?' That lasted about a week. Now I'm not at zero anymore, but it's not that much better. Petal still sniffs me a little bit. She sticks her trunk in my pocket to check for candy. I think she's testing me."

"Have you used the whip?" I asked.

"Not *used* it," he said, "but I grabbed for it once to back Kutenga off. She was in one of her moods. She was throwing stuff into the building, some of them little pebbles, and it ticked me off. So I went for the whip. That she remembered. She took a look at it and backed right down."

We strolled around toward the elephants, Gene's boots shuffling heavily on the rough floor. Now that I was indoors, I could see once again how immense the elephants are, and how terrifying. Gene flicked his cigarette butt toward Petal, and she sucked it up with her trunk and stuffed it into her mouth. Gene smiled. I was horrified. But Gene told me, and Dave Wood later confirmed, that elephants go crazy for tobacco, even in the form of lit cigarettes. He was being nice to the old lady.

Like the other animals, the elephants were indoors because of the cold weather. They had to stay in whenever the temperature dropped below forty. Their ears could get frost-bitten, and they were susceptible to colds. But they hated being cooped up, and right now they looked as if they were working up a case of cabin fever.

With the coming of winter, the true test of Pfeffer's standing with the elephants was approaching. In warm weather, the creatures could go out into the yard to blow off steam. Indoors, the pressure built up and turned their relationship with the keepers into a battle of wills. For the keepers needed to push the elephants into a space where they didn't want to go, and they needed to keep the elephants there no matter how high their tempers rose.

"They've just been out for a couple of hours," Gene said. "So they're feeling better. But if they have to stay in all day, they get rammy—they get upset easily, their attitude changes. On Saturday afternoon, they went crazy because they had to stay in. They looked like they wanted to kill somebody. Inside, there's no freedom, people are hollering and screaming all the time. It's the worst with kids, 'cause they've never heard an echo before, so they're always whooping it up. Drives the elephants nuts. If you can handle 'em when they're all revved up, then you've got it made."

Just then, the Asiatic elephant Peggy, normally bashful, flicked some water from her basin over at Gene. Gene lit into her. "Hey!" he yelled, "Cut it out! I want you to *knock that crap off!*"

Peggy tossed a few more drops our way. *"Did you hear what I said?"*

Peggy stopped tossing the water. She stood in her cage, swinging her trunk.

If he let Peggy get away with that, he figured, she'd try for something more next time. He turned back to me.

"How's Chuck doing?" I asked. "Is there any hope for him with the elephants?"

He gave me an indulgent look. "Let's just say that there are days when Chuck can do it, and days when he can't. He's had his chance. The push is for me to get training, not him."

"Is there anything he's doing wrong?"

"I don't know. I think it may be the way he moves, the

way he talks. They might think he's afraid of them. You can't let 'em know you're afraid. They're like dogs. If they think they've got the edge, they'll take it. There's danger all over the zoo, don't get me wrong. But these guys here"—he gestured toward the elephants—"are so big and powerful. I don't think they even know how strong they are. They've probably never been pushed to the limit and had to use it all. Every once in a while they get going out in the yard. You have to see it to know what I'm talking about. I used to hear about them getting revved, back before I worked in here, and I laughed. I didn't believe it. But now I've seen it, and it's a scary thing." He looked at me for emphasis. "It's a very scary thing."

"You don't have any books to study about this, do you?" I knew it was unlikely.

"Books? Nah, you can't learn this stuff from books." He looked back toward the cages. "No book's going to help you in there. You learn by watching 'em. I watch 'em, and they watch me. See?" He looked over at Peggy, who appeared to be look-ing idly at the wall, lost in thought. "She's watching me right now. It looks like she's not, but she is. They're all watching me. They're waiting for me to make a mistake."

"What kind of mistake?"

"Pushing 'em too far. See, I can make 'em back up three steps, but I'm not going to make 'em back up four. If I try to push 'em four, I'm gonna lose."

"How about Jimmy and Danny? Is that true for them?"

"Hell, Jimmy and Danny could push 'em the whole length of the yard. Jimmy particularly. He's the bull elephant, the head man, he's it. He's God! The elephants would do anything he said. They'd get down on their knees if he asked 'em to. If he yells at me, they're going to go after me."

"When will you know if you've got their respect?"

"I don't know. When I can work around 'em. When I can calm 'em down after they freak."

"When will that be—any idea?" Jimmy appeared in the doorway and looked over at Gene, but said nothing. "I gotta go," said Gene. "The day will come. Maybe I'll have to do bat-tle with 'em this winter. I don't know. But the day will come.

'Cause I'm planning to make this my career. I'll be here till I retire."

He flicked another butt through the bars for Petal, who slurped it up. Then he slipped between the bars himself and hurried through the elephants' cage to see what Jimmy wanted.

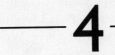

4

Matilda, a camel with a Beatle haircut like her fellow dromedary, Bart, had refused to come inside all last winter, and the
vets were determined not to let her repeat her foolishness
this year.

She and Bart had a small hut in the middle of their exhibit along the walkway by Solitude. Bart had been quite sen-
sible. As soon as it got chilly outside, he went indoors, ambling
along very nicely in that nearly royal gait of his. Most animals
(to be fair, not all) have the sense to come in when it's cold.
The cheetahs, barasingha deer, tahrs, llamas, guanacos, and
chamois all knew when to come into their winter quarters. The
gibbons, under the monorail, had a special warming hut in
their exhibit, and they used it.

Not Matilda. She had simply and categorically refused to
set foot inside her hut. Not that the keepers hadn't tried to
persuade her. Eight of them ganged up on her, four on the
right side, four on the left, and tried to ram her in. Despite
her stouthearted resistance, they managed to get her to the
doorway of her hut, but they fell short of the goal when she
braced her hind legs against the sides of the door and refused
to budge. So she stayed out. Everyone figured that when the
bad weather set in, she would come to her senses. Tempera-

269

tures dropped, she stayed out; it snowed, she stayed out; it blizzarded, she stayed out. She must have been miserable. Snow and ice clung to her shaggy hide like lint to Velcro, and it mounted up. "Most of the winter," said Eileen, "she looked like a walking igloo." But she stayed out.

No one knew exactly why Matilda was behaving like this, although theories abounded. Some people attributed it to a five-year-old's youthful naïveté. Possibly she was unaware of just how bad winter could be. But Dave Wood attributed it to the general stupidity of the species. "They don't do a whole lot except eat and walk," he said. "Sometimes they try to breed."

This last point led to the prevailing theory that possibly Bart had tried something with Matilda when she was last in the hut that had given the place bad associations. Bart was a randy old soul, and he was always trying to have his way with her. Male camels communicate their intentions to the females by biting the backs of their legs, to try to get them to drop to the ground so they can pounce on them. Camels copulate lying down. Matilda, however, would have none of it. Whenever he nibbled her back legs, she bolted.

Regardless of Matilda's reasoning, if any, the vets felt that she should be stashed indoors for the season, forcibly if necessary.

It was necessary. Keith and Mike appeared one morning with a posse of keepers and urged Matilda to reconsider. When she held firm once again, they had no choice but to dart her. The vets took the precaution of locking Bart into his side of the camel hut, lest he be tempted to take unfair advantage of Matilda while she was anesthetized. They gave her a pre-anesthetic to slow her down. "She got kinda limp and droopy," said Eileen. In hopes that the camel would drop near the hut so they wouldn't have far to drag her, Mike waited till she had wandered close to the hut before he shot her, using the full dosage. She received a blast of the anesthetic practically at her doorstep, but then she teetered toward the far fence like a drunkard. As she wove about, she nearly toppled into the moat—that would have been a predicament—and finally came to rest at the absolute farthest point from the hut. The keepers groaned. The vets took this opportunity to give Matilda a TB test and to treat a slight abscess on her jaw. Then the keepers tipped her 1,500-pound body over onto a blue tarpaulin, and with a shout of "Heave ho!" began the long haul back to the hut. The trip took about fifteen minutes. Once she was inside, Bart was let out of the hut, whereupon he ogled her prostrate body through the doorway and drooled on her.

"You know," Mike told me afterward, "that was a really wonderful expression of our zoo. So many people pulling together in the same direction." Then he broke into a big smile. I was glad to hear him joking again.

I went out to take a look at Matilda. She had revived by then and was standing up inside her hut. A more intelligent animal might probably have been wondering how in heaven's name she had gotten there. But not Matilda. She had poked her head out through the opened top half of the hut door. With her head exposed to the elements, I don't think she realized she wasn't in fact still outside. I don't think she realized that anything had happened to her at all.

5

Up in his second-floor office at Penrose, bird curator Larry Shelton can usually be found up to his elbows in catalogs, newsletters, booklets, and other detritus of the curatorial profession heaped on his desk. He has a dark beard, a plume of red hair, a fidgety manner, and the general air of a prep-school classics master. He is learned in his abstruse specialty of passerine, or perching, birds; favors corduroy jackets, loves opera, and has an explosive sense of humor, frequently bursting forth in a crackling laugh that sounds like severe engine trouble.

In mid-December he was all abuzz as he waited for a superb fruit dove to arrive from the Miami Zoo. He tried to reassure himself by noting that after about three thousand shipments in eight years, he had yet to lose a bird—and this included the two blue-faced parrot finches that were left untended for two days at the Los Angeles airport because somebody forgot to load them onto the plane.

"Do you always ship the birds by plane?" I asked.

"Oh yes," he replied in his deep bass voice, with a trace of his Tennessee origins still showing. "Of *course* they fly." He chuckled at the joke. "For a short trip like Miami–Philadelphia, you put them in as small a cage as possible, so the bird

273

has less of a chance of banging around inside and injuring itself. We always pad the top, because one of the big dangers is that the bird will get scared and try to fly up, and bang its head on the ceiling. Still, it's obviously traumatic to be placed in a tiny space in the semidark, and then tossed into the belly of a plane where it is pitch-black. It is a reflection on how adaptable most of these creatures are that the survival rate is as good as it is."

But no matter how much he tried to reassure himself, he was still anxious about the fruit dove, and he grew more so as time went by. However, that gave us a chance to chat, since he was too distracted to do much else.

Larry had always wanted to be an ornithologist. From an early age he had collected the small colorful birds, like finches and canaries, that are called soft-bills. His family's interest in animals, however, had extended only to horses. As a high-school student, he wrote to Dr. James Peters, the author of the multi-volume *Checklist of Birds of the World*, about his chances of getting a job in the field when he graduated. Remarkably, Dr. Peters answered the letter, and said that there were virtually no jobs in ornithology. "That was the part of the letter my parents seized on," Larry said. So he enrolled in the University of Tennessee, majoring in comparative literature.

When he graduated, he moved to Manhattan and started a modest publishing company called Employee Communications, which published special-interest magazines for employees. He also kept about a hundred birds in his apartment. "I had a large living room, and I had the aviaries on one side. They were all lit dramatically, with nice planting around them. It was like having a garden in my apartment."

Happily, his neighbors didn't mind. "I was always careful not to get birds with loud voices. Loud colors, but not loud voices." Nor did the birds ever escape. "Oh Lord no," Larry exclaimed. "One never feeds or cleans a birdcage with open windows. That's basic common sense."

After a few years in publishing, he decided to sell his birds and move to the West Coast. He had sold some of the birds to Gus Griswold, then the Philadelphia Zoo's curator. Griswold was retiring, and he suggested that Larry apply for the job. "I did it kind of on a lark," Larry said. "I didn't think anything of it. But this is as far west as I got. Maybe I should have gotten a better travel agent." He laughed, producing a loud and extraordinary crackle.

Even without further training, he flourished at the zoo, achieving breeding firsts with the hooded pitta and orange-billed euphonia, and others. Larry had developed an almost uncanny ability, as he put it, "to think like a bird." He had enjoyed great success with the purple honeycreepers in Jungle Bird Walk, the zoo's tropical bird house, largely because he realized that they needed privacy. "The birds are highly terri-torial," he said. "There was room for two pairs of the honey-creepers in Jungle Bird Walk, but I suspected I was never going to get any breeding if I had that many because the birds would spend all their time bickering. Those birds eat and bicker. That's all they do. If they are bickering all the time, they won't have any energy left for breeding. So I left one pair in there, and they produced young. I had to be sure to take the young be-fore they left the nest and hand-raise them, or the parents would do them in. They would be competition just like any other bird. Their general feeling is: They have done their duty just to get them to the street.

"It was a little different with the scarlet-breasted tanagers. They needed another pair to stimulate their own breeding. But once they started breeding, they always tried to beat the sec-ond pair up. So we had to leave the two paris together till we got some breeding, then pull the second pair and leave the lovebirds alone."

In 1983 he had had a terrible fright in Papua New Guinea, where he had gone on a collecting trip for the zoo. He was particularly interested in birds of paradise, rare, resplendent creatures with colorful plumes that sweep from their heads like long, feathery horns. May was supposed to be the dry season, but the torrential rains had continued without let-up. Larry and a colleague set up long nets of fine nylon on a ridge to catch the birds as they flew by. Finally, one afternoon a sought-after bird of paradise flitted into the net. The bird became deeply entangled in the fine hairs of the net, and Larry and the colleague were so completely absorbed in disentangling it that they didn't notice that twenty New Guinea tribesmen had silently crept out of nowhere to converge on them. They were naked except for loincloths, and they carried spears and stone axes. They looked at Larry's red hair and his colleague's blond hair with an enthusiasm that the two westerners found alarming. "They were very curious about the colors," said Larry. They also expressed a desire for the bird, which New Guineans commonly used for ornamentation or for eating. "They kept pointing at the bird, and we kept saying, 'No, no, no!' " The tribesmen spoke pidgin English, and the word for good-bye sounds like "afternoon" as pronounced by a parrot: "happynoon." Larry repeated the word over and over as the two westerners retreated, bird in hand, down the hill to their camp.

Because of Larry Shelton's dedication, and his successes, the zoo world turned to him when Bob Beck, a young wildlife biologist from Guam, a small American island in the Pacific, came to a national zoo conference with a serious problem. The birdlife on Guam was mysteriously disappearing, and if nothing was done, six colorful species of birds that lived only on Guam would vanish forever.

Zoos get involved in these emergency rescue operations from time to time. It is thanks to the Bronx Zoo that the American plains still have any American bison. That zoo had salvaged a breeding stock by the time the original herds were destroyed. In 1900 in England, the duke of Bedford was able to save the Père David's deer by commissioning the monk who had first brought them to the attention of the West to smuggle a few from the Chinese palace that possessed the world's only collection. When the Chinese collection was killed off by contagion, the duke of Bedford provided the Chinese with a fresh supply from his private menagerie. Similarly, in 1950 a Hawaiian farmer named Harold Shipman, and Sir Peter Scott, founder of England's Wildfowl Trust, saved the nene goose.

In the Guam case, the local officials had no experience in captive management, so Beck had no idea how to preserve the tiny island's six fast-dwindling species: two flycatchers called the rufous-fronted fantail and the Guam broadbill, the whiteeye, the Guam rail, the Micronesian kingfisher, and the Mariana crow. Zoos, on the other hand, make captive management their business. Larry Shelton was appointed the chairman of the hastily organized Guam Rescue Committee, which immediately set out to capture a decent breeding stock of the birds.

Shelton flew to Guam in July 1983, and when he arrived, he was astounded. "I drove out to Bob Beck's house. It looks out on these vast ravines that have probably never been cut back. It's as lush a habitat as you could imagine. We were looking out at it, and he said to me, 'Do you hear anything?' And I realized: There's no sound! Just about everywhere in the world, the first thing you hear is birdsong. Yet here, in the tropics, there was total silence. There was nothing there. Just nothing. Nothing at all."

They scoured the island, and they were appalled at how few specimens they could find. Only four or five white-eyes remained on the whole island. Shelton managed to net one of them. And while twenty male broadbills had survived, the females were all gone, so that species was doomed. It was probably the same story with the fantails, but since males and females looked identical, Larry couldn't be sure. "I'll never forget watching a Guam broadbill flit around," he said. "He was a pretty little bluish bird with a crest that he could raise. And he was just singing away in this jungle that his ancestors had inhabited for a million years. He obviously didn't know it, but he was biologically dead since there were no females for him to mate with. But he was singing away so happily. That was the saddest thing."

The kingfisher, rail, and crow populations were moderately larger, and the rescue committeemen realized that those birds were the only ones for which they could hold out any hope.

What had nearly wiped out Guam's bird life? When Larry first spoke to Bob Beck, no one knew. The Guam authorities had ruled out agricultural pesticides. If they were going to have any effect, it would have been right after the war, when their use was highest. But the population crashed relatively recently. "There had been a slow decline for about twenty years," Larry said. "But it was not on an alarming scale until about five years ago. Then suddenly it was like that." His hand chopped through the air in an imaginary downstroke on a graph. By the time Larry arrived, researchers had started to rule out diseases, too. Then a circumstantial pattern of evidence began to appear, isolating the cause.

"Back in the nineteen-forties, there began to be reported sightings of something called the brown treesnake," Larry explained. "As the frequency of these reports increased, so did reports of birds disappearing. The odd thing was, the brown treesnake is not native to any of those islands. Until it appeared, there were no snakes of any kind on Guam at all. So man must have brought it in. There was some feeling that the air force introduced it—a pregnant snake may have slithered onto a plane from somewhere—but there was no proof. When I was out there, some of the locals were calling the snake the Philippine rat snake. So it may have been deliberately brought in to control rodents. If that is so, they chose about the worst possible snake for the job—especially on an island with birds that have never been exposed to snakes before. For the brown treesnake is one of the most highly specialized bird eaters in the world. It is long and very slim, which allows it to reach out on branches to get birds in their nests even out on the smallest twigs. The biggest museum specimens from native habitat are eight or nine feet. On Guam they reach eleven feet. They are longer, I have to think, because they are so well fed.

"It's true that insular creatures tend to be more naïve. That's why you have tame birds on Galápagos, and things like that. But I'm sure that if a snake population had reached this density anywhere, there would be a severe problem. There is no natural enemy to control this snake at all. That's why they live long, and eat more. Also, they are becoming something of an economic factor on Guam because they are always crawling into transformers and causing brownouts and blackouts. Right now, the Fish and Wildlife people on Guam are trying to learn as much about the snake as they can in hopes of controlling it. It's impossible to eradicate. They want to make sure it keeps from spreading to other islands."

In the meantime, Larry and his fellow committeemen had focused their efforts on the three remaining species. Larry decided that the Guam biologists should work with the rails, since that species is easy to manage in captivity. He told them what size pens to keep them in, and how to pile some brush at one end for them to breed in. Just to be on the safe side, he recommended flying some of the rails back to the United States. "It always scares me to keep all of a species in one place," said

Larry. Otherwise a single typhoon, outbreak of disease, or fire could destroy the species.

Unfortunately, due to a miscalculation on the part of the Guam officials, they obtained only six rails for breeding instead of the twenty or thirty Larry had wanted. When they realized their error a few months later, the bird was gone from the wild.

Nineteen kingfishers were captured, and nine of them were flown to American zoos for breeding. They came on a private Lear jet belonging to ARA, a food service company whose president serves on the Philadelphia Zoo's board. Two of the kingfishers came to Philadelphia. Larry set them up in a secluded exhibit in the Bird House. Judging by their long, delicate beaks, Shelton sensed that the kingfishers required soft wood in which to build their nests. He sent out workmen to scour Fairmount Park for a suitably rotted log. The kingfisher couple shared in the work of excavating a hollow: They alternated flying at the log to hammer at it with their beaks. "That's part of their pair bonding," said Larry.

After delivering some infertile eggs, the female finally produced a chick over the summer. Larry had to try to maintain their natural behavior if zoos had any hope of ever returning the birds to the wild. Most important, he kept them on their diet of live lizards, even though they cost $1.25 apiece, and each bird ate three or four a week. If that instinct was lost, they wouldn't be able to feed themselves in the wild. And it was to the wild that Larry hoped the descendants of these kingfishers would ultimately be restored.

By now, Larry had made three or four increasingly anxious calls to the Bird House to see if the fruit dove had come in from Miami. Finally, he couldn't stand the suspense any longer, and he decided to walk over to the Bird House and wait. On the way, we saw Beth Bahner, who told Larry that the shipment had arrived at the airport, and was en route to the zoo. Larry was visibly relieved.

We hurried on to the Bird House, and entered by the side door. The Bird House is the neoclassical building that fronts Bird Lake. The side door, however, opens onto a large room that serves as the bird infirmary. There was a cacophony of birdsong and a fluttering of wings. Some of the cages were hooded to give the birds a sense of tranquility. Sonny Woerner, the head keeper, came up to Larry with a dead bird that had been found in the Jungle Bird Walk. Sonny couldn't identify it. Larry looked at it for a moment. "It's a yellow-breasted warbler," he said. "It's not one of the zoo's." Larry figured that it had been brought in sick by a visitor in a coat pocket and deposited in the Jungle Bird Walk on the assumption that the zoo would know what to do with it. Used to being hand-fed, it wouldn't know how to function at the zoo. Also, it could be carrying any number of parasites and diseases. "Whoever it was obviously thought this would be more compassionate," said Larry with evident disgust.

I looked around at the caged birds. They were all beautiful specimens—brightly colored finches, curious hanging parrots that, for a reason Larry didn't know, hung upside down in their cages like bats. A scarlet tanager puffed out its chest as I approached. "That's one of our unfortunate successes," said Larry. "He thinks he's human. He's only supposed to do that kind of display with female tanagers, but he does it to people. I keep telling the keepers—don't baby the birds. The last thing we want is for them to get imprinted on humans. But some of them do it anyway. I really drilled this into one keeper. I kept telling her, don't play with the birds. Leave them alone. Then one day, I came in here and found her with a tanager in her hand, it may even have been this one, singing a lullaby and rocking him. Can you *imagine*?"

He used the Bird House phone to call around in hopes that the superb fruit dove had come to the zoo but been misdirected. There was still no sign of it. I slipped out a door to the exhibit area, to steal a look at the Micronesian kingfishers. They were in a corner display that had been shrouded from public view by some shrubbery. Rather than indicate that anything particularly momentous was going on in the exhibit, a sign said only that the cage was being remodeled. I peered through the leaves like a jungle explorer, and after looking intently for a moment, I could pick out the two kingfishers. They were pudgy birds, blue-winged with long, pencil-pointed beaks. The male's belly was rust-colored, the female's white. And, just as Larry had said, they were knocking a hole out of a rotten log that had been propped up in the corner. As I watched, one would flutter into position and dive-bomb the log, whacking a little hole in it; then the other would take over, and give it another whack. And so it went, *whack, whack, whack.*

When I returned to the side room, I found Larry still acting very nervous. He was pacing back and forth in the room, unable to concentrate. "I feel like an expectant father," he said. Finally, the door opened, and Dave Wood came in carrying a tiny wooden crate about the size of a lunch box. "I think this is for you," he said to Larry.

Larry hurried over to him, and took the crate. He quickly twisted a couple of nails that held one endpiece in place to release his prize. As Larry held the crate, Dave reached in to pull out the bird. It was pale green, about the size of a small pigeon. Feathers flew into the air as the bird squirmed in Dave's grasp. "I've got some elephant smell on me," Dave said. He passed the dove to Larry, and it calmed down almost instantly. Larry rubbed the bird's head with his thumb, like a bishop laying his hands on a tiny communicant, and lulled it. Then he began to feel the rest of its body the way a greengrocer might size up a melon, and gave it a wary look in the eye. Finally he placed the bird in an aquarium-sized cage in which it would get acclimated before going on display.

"You almost develop a sixth sense about these things," Larry explained. "You can tell from the brightness of their eyes if they are healthy. If the eyes are at all glazed, you might have problems. Then you check the chest and back. You feel for how fleshy they are. You see if there is any food clinging to their beak. Later you check the stool sample. But this guy looks fine."

He looked into the cage. "Welcome to the Philadelphia Zoo," he cooed. Then he dropped a hood over it.

6

The Species Survival Plan committee in Minneapolis that governs the sex life of most of the world's captive Siberian tigers sent word in early December that George, the Philadelphia Zoo's big male Siberian, could finally have a go with Martha, the eldest of the three Siberian first ladies. When the zoo received the news, keepers opened up the doors between their cages to form a kind of bridal suite for the couple down in the tiger wing of the cavernous old cat house. But Abigail and Dolly continued to be confined to their solitary chambers down the row.

The Siberians were nearly as big as horses, and the bold black stripes on their blood-orange coats were mesmerizing. The tight cages concentrated their energy and made the animals seem all the more alarming. They were always stalking back and forth and growling, their enormous paws thumping the glossy tiles, their thick tails whipping about, their heads slung low from their shoulders like murderers'.

In the keepers' quarters, an instructional guide to tiger stripes helped distinguish the three first lady sisters: for Dolly, the stripe by her left ear curled into a J; for Martha, a backward L; and for Abigail, an inverted and reversed F. The only one of the sisters I could tell apart from the others was Abigail: When she slept, she always draped her tail in her water bowl.

Now that George had entered the picture, however, Martha had changed from her former indomitable self into a sex kitten. While George strolled about his cage, his mind elsewhere, Martha slunk sexily around in hers. She was clearly trying to entice the big fella through the open door and into her boudoir for a visit. She rolled on her back, spread her legs, and emitted a growly purr. George took one look and turned sharply away. Still on her back, Martha started squirming, as if trying to scratch some deep and elusive itch. George gazed blankly at the wall. Now Martha was positively writhing in what had all the earmarks of sexual agony. But George couldn't have cared less. He slumped onto his belly, then rolled lazily onto his side as if he'd had a long day and all he wanted was a nap. Martha decided to be a little more direct. She got up and padded into George's cage and rubbed her muzzle along his flank, then breathed in George's ear, then swung her rear end around so he could get a whiff of her perfume. It didn't work. It just made George mad. He gave out a thunderous roar and flicked her away with a swipe of his enormous paw. Martha strolled back to her cage, stretched out, and yawned. Okay, if that's the way you want it, she seemed to say, I'll wait.

Down the hall, head keeper of the Carnivora House Al Porta was disgusted with George's performance. "I think we're going to change his name to Boy George," he said. Al is a pleasant, gray-haired, talkative fellow in his mid-fifties. He'd grown up around horses, since his uncle ran a carriage company in South Philadelphia in the thirties, and he had come to the zoo after the war. He was making dinner for the binturongs when I came in.

"Hasn't he ever responded?" I asked.

"Oh sure," he answered. "A few times. But he can't hit the home run. You know what I mean? He can't hit the home run. Martha, now she's doing everything a female should do."

"So I've seen."

"And he's gotten up on her once or twice, but he can't get it in. He's that far from her vagina." He put his hands about a foot apart. "He's riding her too high. Oh well. It's to be expected. He's only five. He just came into sexual maturity last winter. We've got a backup female, in case it doesn't work out. That's Abby. Dolly's out of it, because she's too aggressive. All they do is snarl at each other. They figured that Martha was best for George because she's the most passive."

"I'll say."

"Yeah, she doesn't care about personality. She'll stick her butt up in the air for anybody."

Al finished slicing some apples and oranges for the binturongs, and tossed them in with some Zoo Cake and lean beef. Then he added a dead mouse—a full-grown, from the looks of it—much as one puts a Maraschino cherry on a banana split, and then garnished the whole thing with a sprinkling of mealworms. He pushed the pans to the far side of the preparation table, and rubbed his hands together to get the crumbs off. Then he checked the menu for the slow loris, the new "Cloris" that had come in from Duke. The loris got mealworms, crickets, and, every three or four days, a couple of the baby mice called pinkies.

"But George and Martha are nothing compared to the two male lions we had here before Webster," he went on. "They were a pair. They were always jumping on each other's bones. Queer as a coupla two-dollar bills. A school group came in here one time and saw them. The teacher didn't know what to say. I told the guy, 'Just don't tell them they're both male, and then move on to the tigers.' But those two would switch off, one would be on top one day, the other on another. We called them Tutti and Frutti."

"Queens of the jungle?" I asked

He looked at me in wonder. "Yeah, you could say that. Queens of the jungle. Hah!"

When the food was prepared, Al walked back with me to the tigers. Along the way we passed by the empty cage where Satan, a gorgeous black leopard, used to live. "It was sad about that old guy," said Al. "He was twenty-five, maybe twenty-six. He was getting on. He had one bad leg. Finally his bladder went. He just lost all control and he was really stinking up the place. The vets had to put him down."

The snow leopard, Sergei, sprang toward the bars as we passed and snarled. Then the jaguars got into it, raising a din. Pretty soon, the whole place was ringing with the sound of the jungle. A jaguar had killed a young keeper at another zoo a few days before. The man had failed to secure an automatic door before coming into an adjoining cage, Al told me. The jaguar pushed the door open, came in, and mauled the keeper. "That's why I always tell the new keepers, there's one thing to this job. Locks. Check your locks, and then check them again. You can't be too careful." Al lifted up a sleeve and revealed a nasty scar from wrist to elbow. "I learned early. I brushed by a cage too close and a jaguar reached out and slashed me. Served me right, it did. Doesn't bother me. An occupational hazard."

We went on down the tiger cages at the end of the hall. Martha was in George's cage again. She was banging his rear end with her forepaw. "See?" said Al delightedly. "She's goosing him!" Al approached the bars and greeted George with a hello. George responded with a deep growl. And he stared hard at Al. "See that look?" said Al, gesturing. The eyes were dazed and unfocused. "That means they want to kill you. They're mad now, 'cause it's feeding time. They're friendly up till feeding time, then the predator instinct takes over."

We took seats on the wooden viewing stands by the tigers, and Al reminisced. George was bred in Minnesota, he said, but the three sisters were born here. Their parents were Kunda and Zaia, and Al remembered them well. "Kunda was a magnificent animal," he said. "Must have weighed six, six-fifty, and he had great personality. He was raised in the Netherlands Zoo by a German keeper, so he understood German. We used to talk to him in German. '*Achtung!*' we'd yell when we came up. Or we'd say, '*Spissen!*' That was a joke. Yell, '*Spissen*' and he'd squirt a little pee out in the direction of the public."

Suddenly, all four tigers start jumping about as though an electric current were running through the floor tiles. I couldn't imagine what had started it until I saw a window washer coming down the hall carrying a long pole that must have looked to the animals like the vets' blowgun. The three sisters kept at it, but George soon lay back down quietly. "He's a ham," said Al. "He gets all aggressive, but then he drops it and falls asleep."

Finally, the window washer left and the females calmed down, too, except for Martha who rolled over onto her back and started writhing once more and calling out to George in a breathy growl.

"The cats are always acting up," said Al. To demonstrate, he went over to Abigail. "If I purr at her, she'll purr at me," he said. "Watch." He purred loudly, a gargling sound from deep in his throat. Abby eyed him suspiciously, but remained silent. Al tried again. Abby remained silent. "You stupid old cat," he said with a dismissive wave of the hand.

Ray Hance, the slender black keeper who normally worked in the African Plains, had taken a seat in the stands himself. He noticed me and sidled over. A few weeks earlier I had asked him if he'd known the "Cat Man" that Janet Lidle, the Wolf Lady, had mentioned.

"You're the one who was asking me about John, right?" he said. That was the Cat Man's name. I said that was right.

"I have a question for you."

"What's that?"

"Why'd you say he was crazy?"

"Because that's the way he was described to me," I said, thinking fast. Actually, I had thought he was crazy, too. But I didn't believe Ray wanted to hear that. Besides, the zoo was forcing me to redefine my notion of craziness.

"Well, that guy wasn't crazy. He wasn't crazy. John was just interested in cats. He wasn't crazy at all."

"Oh." I didn't know how else to respond. We both looked at the Siberians, who had now turned to look at us.

"Well, that's all I wanted to say," he concluded. "I didn't want you to get the wrong impression." With that, he slid back over to his seat and continued to watch the big cats in silence.

—7—

Later in December, the three wolf sisters were arrayed one afternoon in their usual constellation about Taboo. But instead of napping, their heads were up, their eyes bright and their ears pricked, keenly following the path of the bulldozers shoving earth around to make way for the new picnic grove that was going up nearby as part of K'Bluey's legacy. Janet had taken up her customary position on the stone bench by the spreading yew bush. Now that winter was fully upon us, she was so completely padded that she was unable to lower her arms to her side, and her gloves were so thick that her fingers were splayed.

"H.B. is back to number two," she said as I pulled up. Speaking in muffled tones through a ski mask, she explained that sometime on Thanksgiving Day, H.B., the wolf who had been deposed by the Napoleonic Trillion, had been able to reassert herself. The zoo was closed for the holiday, so Janet couldn't say for sure what had happened. But when she arrived on Friday, she saw that Koshie, the erstwhile number two female, was cowering before H.B. just as she had before Trillion shook up the pecking order. No doubt, H.B. had taken a run at Koshie, and Koshie had folded.

It wasn't surprising that Koshie had been knocked off. With

her striking white coat, she seemed to have the soul of a bleached blonde. She depended on the goodwill of others; she didn't have the guts to fight for herself. And Trillion, now that her alpha status was secure, hadn't felt any need to help her.

After Koshie's quisling behavior, I was glad that she had taken a fall. It was one thing for Trillion to vault to the top; it was an other for Koshie to ride her coattails to become second-in-command. And I was pleased, too, that H.B. had managed to resurrect herself, if only to number two. She struck me as the healthy, good-natured, girl-next-door type. I didn't want to see her crushed at the bottom of the heap.

One point of the hierarchy was to determine which sister would get to mate with Taboo. It didn't matter to the wolves that he was their brother, apparently: Taboo was not taboo. But the zoo, more conscious of the genetic consequences of incest, had taken the precaution of giving Taboo a vasectomy. "He can do everything except impregnate," Janet explained. "That's different from being castrated. When they're castrated they lose all their sex drive. They did that to Pierre, the male African lion. Next thing you knew, his mane fell off." Janet laughed.

For wolves, however, romance isn't necessarily so calculating: The alpha male doesn't always choose the alpha female. And despite Koshie's drop in standing, Janet had noticed that Taboo paid her a fair amount of attention. She had also discerned that Trillion was watching Taboo with the intense absorption of the lovelorn, so it was hard to say just what would happen. As far as Janet could figure, this winter was when it would happen, most likely in February when the females first came into heat.

The wolves have an elaborate courtship ritual, which Janet was rather proud of. "It's not like dogs," she said. "They just trot up and jump on. Courting wolves walk along side by side, rather like a formal promenade. You might think of it as their holding paws. And they nose each other, and lick each other. Then, when it comes time for sex, she turns her tail to the side and says, Okay, I'm ready. I'm willing. If she's not ready, she sits down."

"The male wouldn't force her?"

"Rape isn't possible. In sex, they stand back to back. The

penis locks into the vagina, and the two sometimes do a kind of pirouette together, slowly circling around. Sometimes they twirl for as long as twenty minutes. But wolves don't always do this. They can conceive without it."

Just then, Trillion roamed over to the corner of the exhibit and picked up speed as she went. Finally, she floored it. She was after a squirrel that had wandered into the exhibit. "She's caught one before," said Janet. "She's quick. She almost got a peacock that came in here with the stupid idea that it could do some foraging. But that's nothing compared to Sunbeam. Sunny once caught a starling in midflight." The squirrel made it up a tree. Trillion stared up at it hungrily like the wolf in *Peter and the Wolf*. "Too bad, Trillion," Janet said. "The other wolves have grabbed some pigeons. Normally they just let them go. They can be gentle if they want to be, and they pounce on them just to play with them. As soon as the animal stops struggling, the wolf stops biting. They only eat them to teach them a lesson if they try to get away." Finally, Trillion gave up and returned to her post by Taboo to resume her examination of the bulldozers.

Trillion came into heat in February, just as Janet had expected. She spotted some vaginal bleeding. As if to underscore the point, Trillion was unusually aggressive. She spent half her time squashing H.B., and the other half wooing Taboo. If H.B. acted up in the slightest—"If she so much as howled," said Janet—Trillion would strut over to her, tail up, head high, fangs bared, and emit a fearsome growl. H.B. would wilt immediately. Trillion was especially sensitive to any advances H.B. might make toward her beloved Taboo. If H.B. happened to walk too close to Taboo (I say "happened"—intentionality is hard to determine in these matters), Trillion was up and at her. H.B. and Taboo once strode together for a few paces by the far fence in the course of their separate perambulations of the grounds. Trillion wouldn't stand for it. She raced over, and H.B. dropped to the ground as though hit by buckshot. Trillion loomed over her. Pathetically, H.B. tried to crawl away out of Trillion's sight. But before letting her go, Trillion subjected her to the ultimate indignity of mounting her.

The rest of the time, Trillion went after Taboo, trying to

get his attention. Often she did this quite boldly by grabbing his tail between her jaws and giving it a sharp yank. She may have been remembering what she used to do with her surrogate father, Shy Boy, who would regurgitate when she pulled on his tail. But Taboo generally didn't like it, often taking an angry swipe at her with his paw. But that didn't keep Trillion from trying it again.

As far as Janet could tell, Taboo hadn't come into maturity yet. But for all their fancy sexual displays, some wolves are simply shy about such matters. Shy Boy, used to copulate behind a rock, out of Janet's sight. It was also possible that Taboo *could* do it but just didn't *want* to. "Sex is not that big a thing for a wolf," said Janet, "like it is for people. He may not want to mate with any of them, but just stay friends with them all. You can never tell with a wolf."

Still, if Janet had to bet on it, she would choose Koshie as the true object of Taboo's devotions. "He spends more time with her than any of the other wolves," she noted. Right now, Koshie was off by herself, merrily batting a rock around the exhibit with her forepaws. Maybe it was that very independence of spirit that made her so attractive. "But you have to remember," Janet went on, "that for wolves it's not just sex. It's a relationship. It's deeper. It usually lasts a lifetime. The wolves are perfectly happy without sex. I think people would be too, if they tried it. I think that, when it comes to sex, people have been sold a bill of goods."

8

Over the winter, Bill Donaldson moved into Solitude, the mansion that had been built by William Penn's grandson, John, back in the eighteenth century. Workmen had been renovating it for some time, changing it over from meeting rooms for the art museum and into office quarters for the zoo. When they were finished, horticulturist Chuck Rogers showed me around the place, pointing out the ceilings—plasterwork done by the Adams brothers of Edinburgh—and the period wallpaper and the curious tunnel leading from the house to the next-door kitchen that allowed food to stay hot while it was being transported. It was all quite charming, but the only problem was that nobody wanted to move over there. It was too far from everybody else in the administration building.

Bill Donaldson solved the dilemma by moving there himself. That made everyone happy except his secretary, Dolores, a usually cheery personality who now assumed a very long face out of utter loneliness as she did Donaldson's typing and answered his phone in the anteroom.

Donaldson betrayed no unhappiness over the move, but then he rarely stayed in his office very long. His old office in the administration building had a massive bookcase that held all the peculiar mementoes he had accumulated in his varied

293

career: a firebox given to him in Tacoma, a gnarled pipe from some zoo reconstruction, a plastic tube resembling a whale penis. But all this was gone now, replaced by a pair of sculpted camels on the mantelpiece and an American flag by his desk. It looked like he had gone legitimate.

However, that was an illusion. I asked him one day if the Philadelphia was planning to do anything like the Miami MetroZoo's famous human exhibit in which an actor took up residence in a cage for a few days.

"We might," he said.

"Would you volunteer?"

"Sure—if they found me a nice mate."

Actually, he often mentioned his desire to put an executive board meeting on exhibit. Special microscopes aimed at the participants would enable viewers to see the clusters of microorganisms feasting on these distinguished executives. Every man is a zoo.

The zoo's own board is stocked with heads of local institutions like Roger Hillas, president of Philadelphia's Provident Bank, and with prominent society people like Joan Scott, an heiress to the Sears Roebuck fortune. All of them are well connected and fanatically devoted to the zoo. Scott was nicknamed "Animal" in her youth, she was so devoted to her pets; her BMW license plate reads ZOO.

The board's chairman is a Main Line investment banker, Clarence Zanzinger Wurtz. He is widely known as "Binky," having been so nicknamed at birth by his older brother, who claimed he would have called him "Winky" if he'd been a girl. "If you are Clarence Zanzinger Wurtz, what else are you going to be?" he asks. Binky has the patrician good looks of an Elliot Richardson. I once saw him holding, rather uncomfortably, a lop-eared rabbit for a demonstration at the Children's Zoo while the rest of the board watched, eating hot dogs. Like many of the board members, Binky had grown up on a gentleman's farm where he regularly milked the cows, collected eggs, sheared sheep as a child. But he doesn't claim to have a magic touch with animals. "Birds bite me, cats scratch me, dogs bark at me," he says. "I was flattered that the elephants recognized me. They threw hay at me."

As chairman of the board, Binky stands at the very center

of the zoo's organizational chart as arranged by Bill Donaldson. The chart is an unusual document. Instead of employing
the usual hierarchical format, it is organized in concentric circles. Donaldson occupies the ring just outside Wurtz at the
center, then comes the circle of vice-presidents, then that of
the various department heads, and so on out to the keepers
and parking-lot attendants. Actually, Donaldson wasn't happy
with the scheme, for in his view it still focused the eye unduly
on him. Unfortunately, it was the only pattern that two-dimen-
sional space could afford.

While Donaldson would like to relegate himself to an outer
circle, his influence is felt forcefully throughout the zoo. Early
in his tenure he hired a young Yale behavioral anthropologist
named Mary-Scott Cebul, now vice-president for planning, who
was working at the nearby Manell Center for Primate Research. Bill thought that, as an anthropologist, she might be
able to tell him something about the zoo species he felt was far
too little studied: *Homo sapiens*. When she arrived in 1979, he
suggested she "do her thing" on the African Plains exhibit—
the row of giraffes, zebras, and white rhinos where Dan Maloney
now works—which had just been completed. She diligently
studied the humans' response to the exhibit.

To evaluate what she saw, however, she wanted to talk to
the project developers to find out what they were trying to
accomplish. "I thought that was only fair," she says. Mary-Scott
discovered that, in fact, there were no agreed-upon goals for
the facility. She looked further, and found there were no such
goals for the zoo as a whole. "The more I got talking with
people," she says, "the more I realized that nobody around
here agrees on why this place exists, or what everybody should
be doing, or how somebody's job complements somebody else's
job. There was no agreement on anything. It seemed to me
that a place with a cause ought to have a party line, and there
oughtn't to be this confusion. And there was a terrible schism
between groups. There wasn't an appreciation of education by
the curators, or an appreciation of the curatorial challenges by
the vet people, or an appreciation of fund-raising by the curators. It was as if nobody liked anybody, and it was all awful."

She reported all this to Donaldson, who listened with his
usual aplomb and told her that the zoo obviously needed to

clarify its mission, and, well, maybe that's what she should do next. She spent the next year talking to everyone at the zoo, and at the end she came up with a twenty-seven page document entitled "Master Plan 1981–1986" that identified the zoo's underlying purpose and outlined its goals for the next five years. Building on the zoo's intentions "for the instruction and recreation of the people" that Dr. Camac and his scientific breathren had specified one hundred years earlier, Mary-Scott was a great deal more specific and down-home in her ideas. The report declared that the zoo was supposed to: provide "a pleasant setting" for viewing exotic animals, gather scientific information, enhance the public's appreciation of wildlife, help preserve endangered species, do research, present a balanced picture of ecology, and preserve Philadelphia's culture. And in order to perform this mission, the report named twenty-one projects, ranging from the new primate center to revised graphics for the zoo's signs and publications. As an accompanying map showed, the projects were sprinkled across the entire zoo. Essentially the master plan said that the zoo needed to be redone right across the board.

By 1985 nearly all the goals had been accomplished. As Donaldson observed, the one mistake the zoo's report made was that it had been insufficiently ambitious. Donaldson was determined to correct that in the 1986 version. This task consumed the zoo personnel during most of 1985 as different task forces came together to analyze and critique every facet of the zoo from the parking facilities to the pungent animal odor in the Carnivora and Small Mammal houses. For the latter problem, Arlene Kut jokingly devised a couple of aerosol sprays called "Carniban" and "Aard-off," decorated the canisters with stinking lions and aardvarks, and added the caution "This product may mask natural odors necessary for reproductive communication." In the end, the plan called for installing a lightly scented fragrance dispenser near the aardvarks.

On the advice of business manager Rick Biddle, Donaldson once asked a group of students at the Wharton School of Business to examine the zoo's strategic planning procedure as a school project. What the students found amazed them. The concentric circles of the management hierarchy made their heads spin. What was worse, they found that the zoo staff usually broke even those loose bonds to form teams at all management levels to attack certain problems, like the master plan.

"Donaldson strongly believes in the unstructured approach to management," the Wharton students wrote in their report. "To the annoyance of a few of his managers, he engages in substantive conversations with all levels of employees on his daily walk through the garden." They wanted the zoo to stick to Wharton School of Business principles of strict hierarchies, a more precise system of rewards for accomplishment (which they found alarmingly "subjective"), and, above all, tighter controls everywhere. In Exhibit I of the report's appendix, they set forth a graph representing the planning process of business theoretician Peter Lorange. It showed how plans should march smartly from the presidential level past the vice-presidents to the lower-management level, and back up the hierarchy as ideas passed from the initial concept through implementation to final assessment.

When he first read the report, Donaldson felt duly chastened. "I figured they were all so smart, they ought to know," he said. Then he thought a little longer and realized that, if anything, the zoo was still too tight, too regimented. He didn't want more controls, he wanted fewer. "I remember from my days in city government that when you start having all these controls, pretty soon they start to become the organizational focus, not the product. The more controls you have, the less activity you get, and you lose sight of your product. Also, these Wharton fellows have the idea that you can fix things and they'll stay fixed. It doesn't work like that. Things break down."

So Donaldson rejected the report and returned to his own management style—and to his success. As fiscal year 1985 rolled around, the zoo continued to expand. At the beginning of that year, the zoo had signed an agreement with Camden, New Jersey, Philadelphia's sister city across the Delaware River, to manage an aquarium the city was building with money provided by the Campbell Soup Company and its own municipal coffers. Donaldson was particularly excited about the project, not just for the $750,000 it would bring the zoo through 1988, but because it would provide a new challenge to hold the interest of his increasingly ambitious vice-presidents, Rick Biddle, Mary-Scott Cebul, and marketing manager Scott Schultz.

This winter, he was formalizing plans to take on the temporary management of the troubled Franklin Park Zoo in Boston, which had fallen into a state of disrepair exceeding even
that of the Philadelphia Zoo when Donaldson took over in 1979.
Rick Biddle had pushed hard for the Boston zoo project, and
was thrilled it was coming through. "We can move in there
and BAM, we can do something," he said, smacking his hand
for emphasis. And, as if that weren't enough, the zoo was looking at ways in which it could provide management assistance
to the many small local zoos in the Delaware Valley and beyond, which desperately needed a boost. By these means, Donaldson sought to make good on his campaign to restore the
Philadelphia Zoo to its historical place as a truly national zoo.

Bill Donaldson must have been pleased to see the zoo
coming along so well, but he was always so cheery and enthu-
siastic, it was sometimes hard to tell when he was genuinely
excited. One wintry afternoon in February, though, he told me
that a few weeks back he had been introduced as the former
city manager of Cincinnati. The remark bothered him for days.
"I'm not the former nothing," he exclaimed with unusual force.
"I haven't gone downhill, I've gone uphill. This is the best job
I have ever had. I was a better-than-average city manager. I
may have been one of the ten best. But I'm a better zoo man
than city manager. You can do more in a job like this."

"What *do* you do, exactly?" I asked. As we sat together in
his Solitude office, the light angled in low through the high
antique windows. He had his pipe going. He was inclined to
take the long view.

"In geophysics they have this thing called the angle of re-
pose," he began. "It's the shape that the side of a hill takes. All
institutions naturally slide into it. I guess I'm one who keeps it
from happening. Or tries to. See, the problem is that you can't
stay still, because out there in the economy and in the world
everything is changing. Rock may look solid, but little by little
it's breaking down. Oh sure, an institution can sit on change
for a little while, but that can't last. Eventually the change is
going to rip you apart. You've got to give a little here and
there, or the whole thing will crack. I think the skill of management is to see these changes coming and give in to them
just enough. What's the third law of thermodynamics? En-

tropy? If no new energy comes into the system, the whole thing runs down. That's the danger that an institution faces. But most institutions dread that new energy."

Yet that is what Donaldson was. Unlike Trillion out at Wolf Woods, Donaldson wasn't out to assert himself. His strategy was to empower others to assert themselves so that they could ride out changes rather than be crushed by them. When he was city manager of Scottsdale, his director of public works had to have a lung removed, and later needed an oxygen tank at work. Donaldson's staff tried to persuade him to demote the man to the easier job of city engineer. Donaldson sensed that the man might prefer a challenge. He promoted him instead to assistant city manager. "I think that's probably the best thing I ever did," said Donaldson.

Donaldson may be the dominant male, but he doesn't lord it over his subordinates. "The crowd I've got here—Mary-Scott, Rick, Scott—they're the ones who are doing all the work," he said self-effacingly. "They're young and energetic. They are more open to change. They like risk. Me, I'm getting old. To a fat old guy like me, doing nothing is probably more relaxing."

9

The zoo ran on gossip, and of all topics few were as sustaining as the subject of Joe Zapezik's wolf dog. Joe Zapezik worked at the Children's Zoo and was known to walk on the wild side. He'd had some regular dogs, but always longed for the real thing. Because wolf dogs are so ferocious, they are illegal in Pennsylvania, as they are in many states. Joe never said where he got his, but get one he did. He named him Babbo and kept the wolf dog penned in his basement.

Zapezik himself was taciturn with outsiders on the subject, but his fellow keepers didn't share his reticence. I went down to the Children's Zoo one afternoon when Zapezik had the day off to see what I could find out.

The Children's Zoo looks like a dairy farm. It's got a barn and a hay silo and several sheds, all of them painted bright red. There are a few animals in there that you don't find on the average farm, though, such as sea lions and llamas, as well as the expected chickens and cows. A couple of keepers were standing around, and I raised the subject of Zapezik's wolf dog. The two keepers looked at each other and laughed. "Oh, we can't tell you about that," one of them said finally.

"I won't tell anyone that you told me."

That did the trick. The keepers were dying to talk about

Zapezik. I began by saying what I'd heard—that Zapezik had had to call in the ASPCA to get rid of the animal. At that, the two of them burst into laughter once more. Eventually they calmed down enough to talk, or at least one of them did. I'll call him Ed. The other one listened as if to a favorite song. At regular intervals, he'd nod and his eyes would dance with delight.

"It was inevitable the dog would have to be put down," Ed began. "The animal was a time bomb. His name was Babbo. He was three parts wolf and one part German shepherd, but the behavior was all wolf. That dog would howl! He'd wake up the whole neighborhood with his howling. How-ooooooooooooooooo!"

Ed howled remarkably well. The sound echoed around the zoo, a frail, lonesome wail that evoked broad glaciated valleys and snow-capped mountains. Some sheep in a nearby pen turned their heads toward us in a momentary panic: Where's the wolf?

"They kept Babbo in the basement," Ed went on. "Joe couldn't go down there or Babbo would tear him apart. Only his wife could go. She wasn't a threat, you see. Babbo had this dominance thing. He was the alpha male. And Joe's wife was his queen. Babbo loved her. He'd wag his tail when she came near. Loved her! But as soon as he saw Joe, grrrrrrrrrr!

"Joe always talked baby talk to Babbo. You know, 'Hi Babbo, good boy, Babbo.' And I'd laugh, 'cause that dog wanted to rip his head off. He chewed up the front steps. He was a menace! A monster! He'd snap two-by-fours like they were toothpicks. But he wouldn't touch Joe's wife, 'cause he was so in love with her. So when Joe was away, she'd let Babbo out of the basement and into the house. She'd leave a sign on the door in case he came home early. WAIT ON PORCH, it said. The dog was a monstrosity."

"How did Zapezik get onto the dog?" I asked.

"He had a Doberman first. Named General. He had it till it was twelve, thirteen years old. Then he moved on to a mastiff, not a little bull mastiff. A *mastiff*. It must have weighed three hundred pounds. A mastiff is huge. It makes a Saint Bernard look like a puppy. He called it Condor. Then he got Babbo. This dog is a freak of nature.

"I don't know where Joe got him, but he got him as a puppy. He loved to watch that animal grow. He'd feed him all the best meats and grains. He'd get all this raw food and whip it up in the blender or cook it in the microwave. He wanted to see how much that dog could consume. He'd ask me, 'How much do you think Babbo ate for dinner last night?' And I'd say, 'I don't know, Joe, how much?' And he'd say, 'Six pounds.' And he'd give me this big smile. Joe loved it. He'd get the dog's food out, and then sit down at the table himself and have some tea while the dog ate his dinner. He'd sip his tea and smoke his cigarettes and watch the wolf dog eat."

"Now that the dog's grown, what's he do with him?" I asked. "Take him for walks?"

"Are you kiddin' me? Take Babbo for a walk? You can't take Babbo anywhere. Cellar, house, yard. That's it. That's his territory. Babbo won't go anywhere else. He'd freak. Even if he did want to go, you could never take Babbo on a leash. He'd pull you all over the state. You couldn't stop him."

"So what happened?'

"Well, finally Babbo grew up to the point that he attacked Joe's wife, so there was nobody who could deal with him. They got Babbo down into the cellar, and nobody could even get near him. So Joe called in the ASPCA to put him to sleep. That was just this weekend. The ASPCA guy came in and was going to dart him, but all he had was a dart on the end of this little three-foot pole, like Babbo is a Chihuahua. Joe says, 'Where's your dart gun?' And the ASPCA guy says, 'Oh, this'll be fine.' So Joe says, 'Okay.' And they go down the stairs into the cellar. And as soon as they put their feet on the top step, they hear this incredible roaring. And the guy turns white and says, *What the hell is that?'* And Joe says, 'That's Babbo.' The guy goes 'Oh shit.' And they go slowly down the stairs. They get to the door at the bottom, and Joe says, 'Now what?" And the guy says, 'Open it a crack and I'll dart him.' So Joe opens it a teeny little bit, and braces his foot against it, and the guy gets his dart ready. They don't hear anything for a second, and then there's this incredible WHAM against the door. Babbo has charged at it and practically knocked them over. And he's screaming! The guy has never heard anything like it. He says, "I think I'll come back with a dart gun." And Joe goes, "Good idea." And so he comes back later and they go through the whole thing again, but he's able to get a shot at Babbo through the crack, and Babbo goes down, and that's it for Babbo."

"What's next for Zapezik?"

"He says he's going to get another wolf dog."

"You're kidding."

"Nope, he says he was just unlucky. The new one'll probably end up the same way, but Zapezik just loves to see those dogs grow."

—10—

Animals are not tamed by incarceration, obviously. During the Second World War, the London Zoo took the precaution of killing off all its snakes, both poisonous and nonpoisonous, and its spiders, lest a bomb release these characters to threaten an already terrified citizenry. Zoos were not immune to the ravages of war. The Berlin Zoo was completely destroyed by Allied bombs during the war; it had to be restocked by other zoos after the armistice in a zoo version of the Marshall Plan. Afraid that Nazi warplanes might cross the Atlantic during the war to strike the East Coast, the Philadelphia Zoo conducted a survey of its two thousand animals in December of 1941 to determine which ones posed the greatest hazard to the human population. A committee reckoned that thirty-eight animals were lethal. Since the list was drawn up in wintertime, the committee didn't worry about the snakes, assuming they would be "incapacitated by the cold." It targeted many of the residents of the cat house—the tigers, lions, leopards, pumas, and jaguars—but the committee also fingered the zoo's two gorillas, Bamboo and Massa, and, according to the Philadelphia *Record*'s account, ten other unspecified apes and its collection of ten bears. The zoo trained a team of sharpshooters; in case of attack, they were supposed to shoot the animals on sight.

The bombs never came, and the animals survived the war. However, the zoo has retained a team of crack riflemen if there is ever an emergency. Dave Wood now headed it up, along with John Groves and Bill Maloney. Their instructions were similar to the standing orders during the war: If any danger-ous animal ever gets out—shoot it. The guns were kept in a room behind the animal services office in Penrose near the Rare Mammal House's primate collection, in the lion house, and in the underground quarters of Bear Country.

Dave Wood showed me the Penrose arsenal one day: two rifles and two shotguns locked in a forbidding glass case on the wall labeled LOADED—FOR EMERGENCY USE ONLY. Dave opened the case for me. He took down one of the rifles and held the gun diagonally across his body in the present-arms position. It was a long-range .30-06, he told me. "If a big cat got out into the garden, this is what we'd use. It's good for bringing an animal down at a long distance." For closer work, he'd use what he called "the cannon," and he took one down from the rack. It was a 12-gauge pump shotgun with a rifle scope. It was thick and heavy. "The rifle can put a hole in your arm," he said. "But this thing'll blow your arm right off."

"Ever had to use one?" I asked.

"Only for target practice. We've never had an incident where a dangerous animal has actually gotten out. Our first priority is to secure the building. If we can keep the animal in the building, then we'll just anesthetize him. But if he gets out and he's life-threatening, that animal has to be destroyed. See, once an animal gets out, you can't dart 'em, because it takes a while for the drug to take effect, and the animal could have killed somebody by then."

Dave Wood also had his own personal .22 for some rou-tine maintenance work—like shooting squirrels and other ro-dents after hours. It was the gun he'd used on the stray cat carrying toxoplasmosis. For special occasions during the day, he used a crossbow. It was just as effective as a .22, but it didn't make any sound. He'd been using one to hunt the seagull he called "Six-pack Harry" because of the plastic beer-can ring that hung around its neck like a necklace. He suspected the bird was spreading disease among the collection, and he wanted to liquidate the creature. "But I have to be discreet," he said.

"Which animal would you least like to see wandering loose?" I asked.

"The polar bear," he said. "No question. They don't just kill for food, or for security. They kill for fun. They are vicious, vicious animals." He put the guns back in the case, and locked it. "I wouldn't like to see a polar bear come up the walkway at me."

That naturally aroused my curiosity, and I set up an appointment to pay the polar bears a visit that afternoon.

The polar bears are the true lords of the zoo. The lions may sport a regal mane, but they are such sleepyheads. Elephants may be fascinating, but they are too self-absorbed to reign. And the gorillas seem too much like unkempt humans. But the polar bears are something else again, with their thick mother-of-pearl coats, their charcoal eyes, and their stately dispositions.

The polar bear exhibit in Bear Country, across from the flamingo yard not far from Solitude, is well designed to show the animals at rest on their sunny boulders and, through large Plexiglas windows, in action underwater, where they would plunge regularly during the warmer months. I could watch them forever as they swam in graceful slow motion, their fur tossed this way and that as their limbs scissored through the deep water.

Now that winter had come, the polar bears seemed more majestic than ever. The humans might hurry along the walkways, clutching their heavy coats tightly to their bodies and cursing the cold. The other animals might shiver with each blast of frosty wind. But the polar bears held their rocky promontory with renewed dignity. Around them stretched a bleak and frigid landscape, streaked with crusty patches of snow. But they regarded it all with icy indifference.

I pounded at the door to the keeper's quarters. Pat Dougherty ushered me inside, and then slammed the door behind me to keep out the freezing drafts of wind. He sat me by the electric heater that was struggling to warm the dank subterranean space as he finished up some scrubbing. Pat is Irish-born, and he speaks with a rumbly brogue I had trouble understanding. He is gray-haired and slender, with an Irish gaiety in his

eyes. Now in his early sixties, Pat had come to the zoo from
England almost exactly twenty years ago. He had spent eight
years in the Bird House, three at African Plains, and two with
the elephants before coming here to Bear Country.

"Glad to be away from the elephants?" I asked.

"They didn't bother me any," he said. "They give a sign
when they're angry and push their ears forward. I learned to
stay away from them then."

"But polar bears are no picnic," I said.

"No, no picnic," he agreed, shaking his head to underscore the point. "The polar bears are dangerous bears. They'll
kill you for sport."

He hadn't come to Bear Country for the bears. He had
come for the days off that went with the job. Keepers with the
greatest seniority get to select their positions in the various
houses in a system called "bumping." Often, particularly among
older-generation workers who regarded a position as more
custodial than zoological, keepers chose jobs largely by their
schedules. The major animal-related factor was whether the
job involved handling snakes. If it did, almost nobody wanted
it. Pat liked the Bear Country job because it offered two days
off in a row.

Knowing that I had come to get a close look at the polar
bears, Pat took me down a short corridor to the back door that
led out to the exhibit. We passed by a couple of enormous
tanks that could have held the ballast for a submarine. Actually, they served to filter and chlorinate the water for the
polar bear pool in strict accordance with federal regulations.
These bears are the most heavily regulated animals at the zoo.
In fulfillment of the federal Marine Mammal Protection Act, a
statute passed in 1972 largely to protect whales, seals, and other
highly publicized marine animals, the exhibit is regularly examined by Department of Agriculture inspectors to make sure
it is up to code in terms of water purity, chlorination levels,
exhibit size, and other factors.

Since it was nearly quitting time, the polar bears were sniffing around the door, eager for dinner. They were named
Clondyke and Coldilocks, and they were both five years old.
Most zoo animals are more intimidating when seen up close,
but not these two. Soft, pudgy, and white as clouds, they re-

minded me of enlarged and animated versions of the stuffed toys my daughter treasured at home. I'd always assumed that those stuffed polar bears gave children a drastic misimpression, and I always cringed a bit when she curled up with one. But now I thought—maybe polar bears are misunderstood.

Pat set me straight. "They'd go for you the way a cat goes for a mouse. They may look harmless, but they're just toying with you." Just then, the larger one, Clondyke, reached his forepaws up against the door. Stretched out, his body reached well above the doorway and blocked out the light.

"Weighs eight hundred pounds," said Pat. "Coldilocks over there is the little one. She only weighs seven hundred." He laughed at that. "But they're both killers. I've never seen them kill anything, but I know they have. I saw a mallard drop in there once. I don't know what happened to it, but all that was left was a bunch of feathers."

We eyed them awhile as they circled around the door. They were so quiet, it was eerie. There was none of the tigers' self-important roaring, or the elephants' shuffling. Their silence seemed deadly.

For comic relief, Pat let me back to another door that looked out at the spectacled bears. These were pleasant, roly-poly animals from South America, so named because of the black markings around their eyes. "You can go in with these fellas," said Pat, reaching for the latch.

I declined, and he chuckled again. Pat had chopped up an apple for them, and he popped a piece through the bars and into the mouth of the female, Bandit. I was jotting something down in my notebook, and suddenly I felt a sharp set of claws swipe across the back of my hand. My heart stopped dead, and a wave of adrenaline shot through my body. Three white lines ran across the top of my hand, but there was no blood.

It was my first true contact with any of the zoo animals. Always before, when I had touched them they were either under anesthesia, subdued by an expert keeper, or so tame as hardly to notice. *I,* was touching *them;* this time *they* were touching *me.* The spectacled bear was quite gentle and separated from me by some thick iron bars, but a sudden overwhelming fear surged over me nevertheless. I don't think it was a response so much to this animal as to all the ones I had seen over the many months I had been at the zoo.

Zoo animals are terrifying, after all. *They might eat you up.* That's part of the thrill of seeing them. But almost without your knowing it, fear works its way into you, and it tightens the nerves. I suddenly could imagine the stark terror that Robin Silverman must have felt when the Siberian tiger turned on her at the Bronx Zoo, or that Dave Wood must have felt when walloped by Kutenga. It is a tremendous shock to feel the malice of an animal, to sense how powerfully it can assert its will. The encounter left me quaking for a few minutes, and it gave me new admiration for the keepers who live with this fear all their working lives.

We bundled up to go outside to look at the sloth bears on the other side of the exhibit. They slumped around like hoboes in the cold. Pat threw a stick their way to get them to move, but they merely watched the stick fly by and remained where they were. "They're not too active," said Pat.

"That's what they want you to think," I said.

Pat laughed again. "Oh no, I watch myself. You can bet on that. No bear's going to get me."

"Me neither," I said.

The polar bears had come around to watch us from their exhibit. The wind ruffled their fur, but they paid no heed. I was freezing, and I told Pat that I had to get out of the cold.

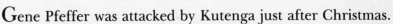

Gene Pfeffer was attacked by Kutenga just after Christmas.

He had been giving the animals hay late that morning in their barred stalls in the Pachyderm House. He'd fed Kutenga, then walked around to feed the others. As he was tending to Petal, Kutenga whipped around.

"Her ears were out and she was comin' for me," he told me over the phone in January as he was finishing his convalescence. "She hit me in the head with her tusk. The blow glanced off, but it knocked me down toward the bars. She hit me again, and this time she knocked me through the bars and out. Then Jimmy came runnin' out, smacked her and she took off. My head, ribs, back, and shoulders were all bruised up pretty badly. Nothin' was broken, but the bruises were bad, real deep. So deep they never went black and blue."

"It hurt?"

"Hell, yeah. But I been hit harder."

"When was that?"

"What do you want, my whole life story?"

"Sorry. How come she went after you?"

"They'd been in all morning. Kutenga just got rammy. It was one of those things."

The fire-department rescue squad had been called to take

SO. ST. PAUL PUBLIC LIBRARY
106 3RD AVE. N.
SO. ST. PAUL, MN 55075

Gene to the hospital. "They thought it was a joke. They kept sayin', 'Ain't that somethin'! Smacked by an elephant!' I told 'em it wasn't funny."

"Has it made you think twice about going back in with them?" I asked.

"I'm not fearful, if that's what you mean. That's never once entered my mind."

"I'd imagine you'd think about it, at least."

"Nope."

I saw Gene at the zoo a couple of days later. He was out in the yard with the elephants, scooping up some boluses into the wheelbarrow as usual. He looked the same as always. Possibly he was more wary, and possibly Kutenga was more attentive toward him, but I didn't notice. It looked to me as though nothing had happened.

"Has the accident set you back with the elephants?" I asked him when he finally stopped for a moment by the fence.

"Nope," he said. "I think I'm right where I was."

—12—

As winters go at the Philadelphia Zoo, this one hadn't been so bad. The snow had been slight, and the days of subfreezing temperatures few. Still, by the time Groundhog Day rolled around, everybody was pretty much sick of the cold and eager to hear that winter would soon be over.

Anyone who doubts a groundhog's ability to foretell the end of winter hasn't spent enough time in Pennsylvania. Pennsylvania is to groundhogs as Kentucky is to race horses. They've got a passel of them, one for every five acres, and they have a resident population, the Pennsylvania Dutch, who possess inflated ideas about the prognostications of hedgehogs in their native Germany. Lacking hedgehogs in America, they have transplanted these personality traits to Pennsylvania's groundhogs, which are known elsewhere in the country, and to Dr. Snyder, as woodchucks.

The association of weather forecasting, groundhogs, and February 2 can be traced back to the early 1880s in the small town of Punxsutawney, eighty-five miles northeast of Pittsburgh, when some sportsmen decided to seek relief from the interminable winter by repairing to the woods to hunt groundhogs. A hibernating species, groundhogs are pretty easy to catch in the winter because they're dead asleep. You just have to fish

313

around in their burrows until you feel something soft, pull it up by the fur, bop it over the head, and you've got dinner for three. These fellows fricasseed a bunch of them, then got stewed themselves on jugs of hard liquor, and, all in all, had such a lovely time they vowed to repeat the episode every winter. More people came along each year, and before long the town had a genuine event on its hands. In the process of dignifying the matter, the alcoholic content of the program was substantially reduced and the winter-ending dimension played up. By 1886 the local newspaper was calling the group the Punxsutawney Groundhog Club and billing the town the weather capital of the world.

Nowadays, they don't eat woodchucks anymore on Groundhog Day in Punxsutawney, and the Groundhog Punch is distressingly alcohol-free. The president of the Groundhog Club raps on the underground bunker of the official ground-hog, nicknamed "Punxsutawney Phil," in a small park called Gobbler Knob. The animal dutifully makes its appearance, and the president converses with it in a secret language called "groundhogese." The club then issues a proclamation about whether or not the groundhog saw its shadow. If it did, it was supposedly so frightened by the sight that it went back into its burrow and prolonged winter for six more weeks; if it didn't see its shadow, it stayed outside and winter was officially over.

The Philadelphia Zoo had been watching the publicity harvest reaped by Punxsutawney Phil with mounting envy, and a few years back decided that since it, too, had a good supply of groundhogs, it should get into the act. (Perhaps out of grat-itude, the zoo later supplied the current Phil with a wife named Philamon, but, groundhogs being shy about breeding in cap-tivity, no issue had been forthcoming.)

Each year, the zoo hauled out one of Dr. Snyder's 150 woodchucks, and enticed Dr. Snyder himself to make an ap-pearance. This year, the event was sponsored by a local radio station, and presided over by a dark-haired, deep-voiced DJ named Ross Brittain. Ross had managed to lure about a hundred people of all ages to a portable, makeshift hollowed-out stump that would pass for the groundhog's temporary winter quar-ters, for what Brittain termed a "coming-out party." The stump was placed by the Reptile House, across the walkway from the camels Matilda and Bart. Matilda, still locked up in her hut for the winter, had a sizable stake in the outcome. Current weather conditions weren't very promising: The bright sun sent deep shadows everywhere.

The stump was equipped with back and front doors; the groundhog was supposed to enter the stump by the back, or stage, door, and emerge by the front into what appeared would be a blaze of sunshine.

For such a scholarly fellow as Dr. Snyder, it must have been a little painful to be involved in this event. Snyder, in fact, had recently returned from a trip to China, where, as a guest of the government, he had helped to set up a study program on the marmot (a cousin of the woodchuck that is plentiful in China) to examine in them the genetic links between hepatitis and liver cancer. The Chinese are particularly susceptible to both diseases. In gratitude, the Chinese government provided Dr. Snyder with a private Mercedes-Benz limousine and a chauffeur for his travels about the country. "I was treated like royalty," he said. That pleased him, but he let it be known that he could best be compensated by a gift of a giant panda— or, better yet, two. The Chinese said they would consider the matter.

Snyder was mildly concerned that the zoo's veteran weather forecaster, Peanuts, was not available for duty this year. He had died last spring at the age of six. Consequently, the zoo had pressed a rookie, Chance, into service. "Somebody said you're taking a chance," he explained, "so that's what we named her."

While sunny, shadow-producing days might seem likely to mean spring was here, Snyder said that in fact cloudy days were apt to be warmer than clear ones, which are often clear merely because a cold-air mass is moving through. Male groundhogs are particularly sensitive to such temperature changes. The higher temperatures trigger their hormones to produce higher testosterone levels, rouse them from their winter-long sleep, and send them forth to do battle with other males for mates. The females generally sleep for another month, waiting to see who won. Chance, however, was female, which may have contributed to Dr. Snyder's anxieties: She wasn't so heat-sensitive. He'd picked her because she didn't mind humans.

His comments were interrupted by eager shouts of "Here she comes!" and the sight of Chance, in a metal cage, being carried out to her stump by Susan Pajkurich, one of Snyder's lab assistants. Chance looked like a furry brown pillow, but she had a buck-toothed face. In the reverse of a star's entry, the lights went out as she made her appearance: Heavy clouds rolled in overhead and plunged us all into darkness, gloom, and, supposedly, springtime.

The event did not begin auspiciously. Susan had trouble evicting Chance from her cage, and had to shove her out the door and cram her into the back of the makeshift burrow. She shuffled inside rather resentfully and then took up a perch just inside the front door. We could all see her hunched there, blinking, her little twitchy nose and overbite poking out into the doorway. It didn't look as if she wanted to come out. "We'll get that sucker out," said a cameraman. Susan sprinkled some slices of apple and banana in front of her to lure her forward into the darkness. "They're her favorites," she explained sheepishly to the viewing audience.

"She looks frightened to death," said someone behind me.

"Why don't you stick a finger out?" asked someone else. Susan didn't smile. In fact, she bore a long scar on her arm from one of many grim encounters with woodchucks over the years.

"Okay, Chance," said Ross Brittain, suddenly growing bored, "it's time to come out now."

Chance didn't budge.

Susan disappeared into Penrose and returned with a secret weapon: a Girl Scout cookie and, with a gloved hand, brought it close to Chance's quivering nose.

"If that doesn't work," said Ross Brittain, "maybe you could detonate a small nuclear device."

It didn't work.

"Hey, Doctor Snyder, you got any female groundhogs around—that'd bring him out!"

Snyder didn't respond. Obviously, female groundhogs

wouldn't help with a lady. And a male wouldn't do much better, since males don't seduce their mates.

Seeing the delay in the proceedings, Dr. Snyder made a
little speech about the origins and significance of Groundhog
Day. But at this point, he was so befuddled by Chance's performance that he got the key point backward, declaring that a
sunny day would end the winter, not a cloudy one.

When he was finished, Susan gave Chance another shot at
her cookie.

"Hey," exclaimed a cameraman, "she's really ramming that
down my man's throat!"

When that failed, sterner measures were required; so Susan equipped herself with a heavy leather glove that went clear
up to her elbow. It reminded me of the falconer's glove that
Elaine Chu had worn to hold K'Bluey.

While Susan suited up, Ross brought out "the ultimate
weapon"—some Hamster and Gerbil Munchy O's and placed
them provocatively before Chance, but she didn't bite.

Susan was ready now, and she marched over, reached in,
hauled Chance out by the scruff of the neck, and put her firmly
on the ground in front of the stump. The clouds had thinned
by now, but I couldn't see any shadows. While I searched, the
TV and still cameras moved forward like advancing artillery,
closing in so tightly as to block out whatever sun there was.
Chance took one look and dashed back into her burrow. This
time Susan didn't bother to pull her out; she pulled the stump
away like a waiter removing a heating bowl from a dish at a
fancy restaurant. Chance sat there, panting. Her whole body
heaved with each breath.

"Does she see her shadow?" someone yelled out.

Just then, the clouds parted and beams of radiant sunlight
shone down on Chance like a spotlight from heaven. Possibly
out of stage fright, Chance took off. In the hobbling way that
woodchucks have of locomoting, she dashed right between the
legs of a startled photographer, and made for a pile of junk in
back of an adjoining building. Susan grabbed a net and dashed
after her. Ann Hess, who had been standing by, joined the
chase with a net as well. The two plunged into the bushes after
the woodchuck.

Dr. Snyder took the stage once more to answer any questions from reporters. As they fired away, I saw Ann and Susan dashing back and forth with their nets high, as if chasing an elusive butterfly. Obviously, they were having a little trouble with the roundup. Finally, the two converged, one net dropped down, then another, and Chance was returned to captivity.

Arlene Kut from PR was asked what the verdict was on the winter. "I'll tell you the truth," she said. "I don't know. Let's say she predicts three more weeks of winter, then three weeks of spring."

As Arlene spoke, Susan carried Chance back in the net and restored her to her cage. We all watched her disappear into Penrose.

A TV reporter then cornered a small boy and asked him whether he believed in the groundhog.

"Yes," he said after a few moments of intense consideration. "I do."

——13——

Well, Arlene must have interpreted Chance's prediction backward. A thaw set in and continued to loosen the ground and muddy the walkways for the next three weeks, and then the cold returned in force.

Up in the accounting office in early February, Rick Biddle was pleased with the warmer weather because it was bring ing in more visitors than normal, and compensating for a slight failure on the part of the TreeHouse to pull in its share of paying guests regardless of the ambient temperatures. "They told me that TreeHouse would be the best thing since sliced bread," he groused, "but I have yet to see it in the figures."

Out in the garden, however, the animals carried on as usual, following that blend of hormones, instinct, and personality that constitutes free will among the lower orders.

In late February the wolf Trillion had the experience of a lifetime. She mated with Taboo. Janet didn't see it, but she knew that it had occurred by the behavior of the two wolves afterward. Taboo was more accepting of Trillion, and the two wolves now spent most of their time together. He didn't even mind when she yanked his tail.

"Any courting?" I asked.

"A lot of sniffing, and striking poses."

"Poses?"

"Yeah, they stand up tall with their heads up." Janet imitated the position. She looked like a Roman emperor. "It's like they're saying, 'Here am I, I'm handsome.' "

"And they do this together?"

"Yeah, it's mutual, one starts, then the other. If they both start doing it, that means they're ready for sex."

She had seen Taboo mount her several times. "He just climbed on," she said. "There weren't any pirouettes. They're too young for that. You have to remember that they're only two. They're still very inexperienced."

"But Trillion's happy."

"Oh yes." Janet nearly swooned as she said it. Then she giggled with delight at herself. "I can tell by the way that she stands there with her eyes closed when Trillion is up on her. It almost looks like she's basking. Her chin is down, her eyes are closed, and her ears are up. That's about the most that a wolf will ever show. You don't want to jump up and down with a wolf on your back." It was the final culmination of the campaign that had begun in December in that terrible fight I witnessed in which Trillion toppled H.B.

"How about you," I said. "Are you pleased?"

"Oh sure. I'm glad there wasn't a lot of nastiness."

Even to my untutored eye, on this March day I could tell that something significant had happened, although I wouldn't have known precisely what. The tension had gone out of the foursome. The wolves looked like teenagers around a pool on a hot day. They might occasionally get up and move around, but for the most part they were content to lie there lazily.

Taboo clearly regarded Trillion as his leading lady. But he also made some time to play with Koshie. Although she was Trillion's age, she hadn't yet come into heat. They would tumble around together and pass bones back and forth. Janet said that Trillion didn't mind the attention her beau was lavishing on Koshie. "She's a nice alpha. She doesn't pick fights. She occasionally gets angry if H.B. gets pushy. But even then it's just one big growl and bark. They haven't gotten into a real fight since that day in November. Small as she is, she can still get mad."

Janet cackled at the memory of Trillion's explosive temper at the start of the winter.

H.B. was lying on a rock in the front of the exhibit. "She's still a little restless," said Janet, "but not nearly as much as before. I wish she'd settle down."

"You don't think she's given up?"

"Maybe not. The dominance order could still change. Over the course of a lifetime, all three females could be dominant. If Trillion fell sick, H.B. could take advantage of it. She might be number one again, who knows? But she'd probably still be Taboo's mate. The wolves are pretty faithful."

Suddenly, H.B.'s ears pricked up and her eyes roved the distance as though she were pondering something.

"She's just looking at the wheelchair over there." Janet pointed to a paraplegic being wheeled off toward the next exhibit.

I asked her how her short story was coming, and she laughed again. She hadn't done any work on it in the last few years. It looked like that tale would remain unfinished. She was too busy here with the wolves during the day, and with her work on the *WOLF!* newsletter the rest of the time. Her latest issue was on the question of wolf hybrids. She'd heard about Zapezik's wolf dog, and she wasn't surprised by the outcome. "What could he expect?" she said. "They're wolves. They still have their predatory instincts." She herself, for all her love of wolves, would never actually want to own one. She contented herself with knitting scarves out of wolf fur culled from the exhibit for her by the keepers, writing about wolves, looking at the pictures of them strewn over the walls of her house, and visiting these wolves in the Philadelphia Zoo.

She was sorry to see the winter draw to a close. The wolves were most active in the cold, and there were fewer people around to bother them, and her. Come Easter, just a few weeks off now, the new visitor season would start up again, and she wasn't looking forward to it. "From Easter through the first week of June," she said, "that's the worst. That's when you get the school groups." She could see it all now: the incessant howling, jostling, and running about of children from what she called "Banshee Elementary, Cacophony Middle

School, and Boom Box High." No, she wasn't looking forward to it at all. But she would return to her wolves, five
days a week, through the spring, and all the seasons that followed. Why?

"I want to see what happens," she said.

The courtship ritual of the Indian rhinos Billy and Xavira. A male rhino can ejaculate nearly once a minute for an hour and a half. *Lauren J. Lewis / Asterisk Inc.*

Xavira getting acquainted with her baby, B.J., not yet one day old, in her stall inside the Pachyderm House. B.J. was only the ninth Indian rhino to be born in captivity in the U.S. *John Sedgwick*

Dr. Wilbur Amand feeding a giraffe *The Zoological Society of Philadelphia*

The polar bears Coldilocks, with her face to the camera, and Clondyke. Despite their cuddly appearance, they are probably the most dangerous animals at the zoo. *The Zoological Society of Philadelphia*

The Siberian tiger George taking a drink in his rocky outdoor exhibit outside the Carnivora House *The Zoological Society of Philadelphia*

Dr. Robert Snyder, director of the Penrose Laboratory, holding his favorite woodchuck, Peanut, the late, great weather forecaster *The Zoological Society of Philadelphia*

The Wolf Lady, Janet Lidle, in warm-weather attire near her usual perch overlooking Wolf Woods. Trillion and Taboo are visible in the background by the fence. *John Sedgwick*

Trillion, foreground, and Taboo share an intimate moment. *Janet Lidle*

Trillion showing Koshie who's boss *Janet Lidle*

Samantha looking a little grumpy as her baby, Chaka, yawns *The Zoological Society of Philadelphia*

Sam samples a leaf at the new World of Primates exhibit while Chaka takes in the scenery. *The Zoological Society of Philadelphia*

It took five men to carry John down the ramp to the new World of Primates exhibit. From left: Bob Berghaier, Bill Maloney, George Konopka, Charlie Fagan, and Mike Barrie *Anne Bettendorf/Asterisk Inc.*

Roseann Giambro comforting Anaka while her mother, Snickers, is examined in the Penrose Annex *Anne Bettendorf/ Asterisk Inc.*

Zoo dentist Carl Tinkelman, left, and curator Karl Krantz, settling Jessica down onto the examining table at the Penrose Annex *Daniel Troy/Asterisk Inc.*

Wilbur the drill yawning on his island at the World of Primates *Terry McBride/ Asterisk Inc.*

Spring

——|——

In early April, buds appeared on the fruit trees around Bird Lake, a chorus of warblers returned to the zoo, the crocuses popped up on the fringes of the lawns, and the new primate center on which Rick Biddle pinned so many of his hopes for the zoo's financial future lurched toward its June 1 completion date.

Or so everyone thought. The site by the Tiger Terrace cafeteria still didn't look like much, just a big mud pit with a half-finished building on one side, and that caused some concern. Biddle strode about the site more purposefully than ever, and Bill Donaldson began issuing uncharacteristically stern pronouncements like "*Something's* going to open on June first or else."

The exhibit was intended as a kind of primate Shangri-La, a naturalistic retreat on four moated islands, with two support buildings along one edge of the site. Back in the fall, the Caterpillar tractors had scooped out the moats and carved out the islands where the gorillas, drills, orangutans, and gibbons would romp. Over the winter, the workmen had worked on the buildings' interiors. The larger one, newly constructed, would house the animal "bedrooms," as they are called, keeper workspaces, and breezeway viewing areas looking through to the open-air exhibit. The other was the old Hewitt-Furness Kangaroo House; it was being renovated to hold a few explanatory displays, a food-preparation area, and an exhibit of tiny Geoffrey's marmosets.

Now, in this final push starting in April, the elaborate plumbing for the moats would be installed, some electrical restraint wires laid down, hundreds of trees and bushes planted, the buildings polished up, the viewing areas constructed around the exhibit, thousands of finishing touches applied, and most important of all, the center's balky primate residents moved in.

It had been a long five years for the project. The idea was hatched when the zoo underwent its first round of self-criticism inspired by Mary-Scott Cebul in 1980, and no one had a good word to say about the dreary old Monkey House where Massa had spent so many of his days. When Bill Donaldson first toured the zoo in 1979, he had taken the Monkey House as emblematic of the zoo's woefully dilapidated condition. Designed in 1907 in an ecclesiastical style by the University of Pennsylvania's Theodophilus North, the building had treated its animal exhibits as objects of nearly religious devotion. The animals' cages were ringed around the walls, and they were drenched with light from the high clerestory windows. By the 1970s, however, the inmates were fallen angels. Terracotta tile was sliding off the roof, and rainwater pooled on the floor. When Venturi, Rauch and Scott Brown, the primate center's architects, first looked into the project, they briefly considered restoring the building, as they had the wonderful old Furness-Hewitt Antelope House to make the TreeHouse, but they realized it would cost at least $4 million. Out of a $6 million budget, that wouldn't leave much for the open-air portion of the exhibit, which was assumed to be the project's basic thrust. It wasn't much to look at, anyway. In 1983 the building was razed, and construction of the new primate center was begun.

In selecting a designer for the center, Bill Donaldson might logically have turned to a firm that specialized in zoo architecture, such as Jones and Jones of Seattle or Zoo Plan in Kansas City, both of which had designed a number of exhibits and a few whole zoos. But he was leery of off-the-shelf architecture. He figured someone new to the trade might bring a fresh vision. This squared with his philosophy of lighting fires under people and then waiting to see if they'd burn up or take off like rockets. Plus, he repeatedly said, he wanted to find someone whose work would still be considered significant in a hundred years. That was hard to imagine in the case of a standard zoo architect. Besides, he had liked Venturi, Rauch and Scott Brown's work on the TreeHouse, and also recognized that the firm had developed good working relationships with such zoo principals as Mary-Scott Cebul; so he decided to keep them on for this new project as well.

Like so many things that Donaldson did, this was also a little crazy. Besides having no zoo buildings in its extensive portfolio, Venturi, Rauch possessed a reputation for wandering a little farther out than Philadelphians were accustomed to go. Lead partner Robert Venturi spoke apocalyptically of the need for complexity and contradiction in architecture, and toured Las Vegas for inspiration. Critics charged that his most seminal work, a private house he had built for his mother in nearby Chestnut Hill, might have been designed by a child, what with its oversized square windows, imposing chimney, and standard inverted-V roof. Far from being distressed, Venturi expressed pleasure at the notion. "I like to think this is so," he wrote, adding mystically, "that it achieves another essence, that of the genre that is house and is elemental."

Venturi himself had a hand in designing the primate center, but the bulk of the work fell, in 1981, to a twenty-four-year-old Alabaman named James Bradberry who at that point had designed a museum, some expensive houses, and a few condominiums. While he did possess some skills at preservation and an interest in landscape architecture, he had never done anything like the primate center at all. Few people had.

Bradberry spent most of his days laboring over the blueprints of the primate center at a long table at the Venturi, Rauch headquarters. The offices are located inside a rehabbed warehouse in the revived Philadelphia section of Manayunk, up the Schuylkill from the zoo. Bradberry is handsome and mod; he has a heavy-on-top, slim-on-the-sides haircut and wears Armani ties. When I went out to see him in the fall, he had been working on the project for four years. "I feel like one of those guys that built the thirteenth-century cathedrals," he said wryly in his slight southern accent.

At least those builders knew what a cathedral was. It was hard to define this "primate center." No single analogy explained it. It was part apartment building, part prison, part garden, part museum, part hospital, part animal cafeteria. And while primates had been displayed in the open air before, never had so many different species been combined in such a small (1.1 acres) exhibit, and only rarely were they shown in a natural, rather than synthetic, environment of trees, shrubs, and grass.

Unfortunately for Bradberry, each of the exhibit's separate functions created a separate client group to pester the designer. That was exhausting. Normally, an architect has one boss—the owner or, for larger projects, the developer. Here, Bradberry had any number of bosses. He had Mary-Scott Cebul, who articulated the zoo's vision of what the exhibit should be; Rick Biddle, who pressured him to keep the price down; the keepers, who pushed him to make the building easy for them to use; Wilbur Amand, who spoke for the animals; the maintenance crew; the vets; the visitors (not that, mercifully, he had to deal directly with any of them); and of course he had the residents—the eight species of primates who would live at the center—and he spent a fair amount of time watching them to see what they were like. He had hardly seen a monkey before he started; he now knew dozens of apes by name. He had been particularly drawn to Massa, and, like many people, often went to visit the old gorilla when the inevitable stresses built up.

Of the primate center's many functions, its prison aspect weighed most heavily on Bradberry's mind, since it was probably the most fundamental. It was also inherently the most complicated, and made all the more so by the need for the exhibit to look "natural." As Mary-Scott Cebul put it, "We're designing Alcatraz and making it look like Covent Garden." In the case of human prisons, it is pretty well established what barriers are necessary to restrain the prisoners; it is clear how high they can jump, how far they can leap, what they can punch through with their bare hands. With other primates, these parameters were by no means clear. How far should a branch of a tree be from the outer wall? How wide the moat? How thick the glass? Heini Hediger's *Man and Animal in the Zoo,* and Alison Jolly's *Evolution of Primate Behavior* offered some insights, but opinion varied widely, and no one knew what an animal would do if it became deeply stressed. Bradberry was pretty sure that orangutans didn't swim, for example, but he wasn't completely sure. And so the moat couldn't be relied on as a restraint. Nevertheless, it was generally agreed that primates could do things that would scare the bejesus out of you. Most other large mammals are confined to the ground, but primates can jump to great heights and climb to still higher ones. They have nimble fingers and an opposable thumb and they are frighteningly strong—particularly the larger apes. "They can do things that no other animal can do," said Bradberry, "like remove a bolt head, or undo a lock, or shear off a galvanized screw. Also, they're real persistent, they'll keep picking at something, and picking at it. That's why it's so important for the zoo to give the animals things like raisins in the grass to hunt for. You've got to keep 'em distracted."

In planning the broader concept of the site, Bradberry toured other zoos and looked carefully around the other exhibits here in the Philadelphia. If there was one thing he hated, it was the big Gunite rocks that so many architects were throwing up in the name of naturalism. He thought a particular disaster was the current Rare Mammal House, which put the gorillas in a yard backed by a rocky outcropping modeled on the Pyrenees. "The whole idea of making buildings look like fake rocks is just hogwash," Bradberry said. "It's done to be naturalistic, but primates don't come from rocky terrain. I think we should let buildings be buildings, and let exhibits be exhibits."

But what kind of building and what kind of exhibit?

Bradberry himself was drawn to the Pachyderm House that Paul Cret had designed for the zoo in 1941, the one like a massive Pennsylvania Dutch barn. "To me, that's one of the great buildings of all time. The scale of that building as related to an elephant is just so fabulous. Architects often used whatever existing architecture was most closely related to the exhibit they were building. That's why the old Monkey House was like a church. It solved a particular question of lighting. In the case of the barn, the massive scale seemed to be related to the massive size of the elephants."

As Jim pondered the problem, sketched designs at his desk, tossed them out, agonized, and brainstormed with other architects, he ended up recapitulating much of the history of the field. Zoo architecture has so little continuity, so little tradition, that designers inevitably must decide for themselves the basic point of it all. At bottom, they always strike the fundamental incongruity of the task: Animals don't belong in buildings.

But in his primate center, Bradberry hoped to release zoo architects' historic hold over captive animals. The first buildings were created for the royal menageries, and their design inevitably reinforced the idea that these creatures were subjects of the Crown. The architecture itself was royal, full of flourishes and high-flown ornament. For example, at Empress Maria Theresa's menagerie, built in 1752 at Schönbrunn Castle outside Vienna, the exhibits maintained the rococo style of the palace. At the empress's direction, the displays were clustered in a circle about a gazebolike dining pavilion so that she could gaze out upon her collection of camels, elephants, and zebras while she sat at breakfast.

As these menageries evolved into zoological gardens, the sense of royal dominion receded and the architecture began to reflect the captive beasts' exotic nature. In 1860 the Cologne Zoo exhibited its ostriches in a mosque, and its elephant house displayed a strong Moorish influence. At the 1843 Antwerp Zoo, the elephant house was designed as an Egyptian temple, its columns covered with hieroglyphics. And the Berlin Zoo, one of the world's most lovely animal gardens, built an elephant house in the 1870s in the form of a Hindu temple, with fezlike domes and dramatic stone columns in the interior that were capped with carved elephant heads. And its elliptical giraffe house was decorated with minarets. All of it, sadly, was reduced to rubble by Allied bombs in the Second World War.

Furness and Hewitt's work at the Philadelphia Zoo in the 1870s was shaped by the same philosophy. Its flamboyant gingerbread architecture evoked the enchantment of the strange beasts that creep through dark jungles. The attitude held at the Bronx Zoo as well; the neoclassical architecture in Astor court was festooned with sculpted elephant heads, lions, and crawling monkeys.

* * *

At the turn of the century, however, Carl Hagenbeck, the animal collector and zoo man, took this tendency several steps further. Why be content merely to *suggest* the animals' native lands through architecture? Why not *create* the lands themselves? With this thought, Hagenbeck invented the naturalistic style so widely in evidence today.

One of the great figures in zoo history, Hagenbeck looked a bit like a New England ship captain. He had a weatherbeaten face and sported a stubby beard called a chin fringe. He was born in Hamburg, Germany, in 1844, the son of a fishmonger. Some live seals turned up in a supplier's fishing nets one day, and the elder Hagenbeck decided to exhibit them for a fee in large wooden tubs outside his house. When the seals drew a crowd, he moved on to other animals, and the family trade was begun. By age ten, Carl was pressed into service in gathering these creatures from German dealers. At fifteen, he took over the business entirely. Besides exhibiting the animals in a small menagerie, he also got into the trade of collecting and trading them; the American circus impresarios P. T. Barnum and Adam Forepaugh were among his clients.

Hagenbeck was always fussing with the idea of what a zoo should be, and one of his first brainstorms was to import not only the animals of these distant lands, but some of the resident people as well. He termed them "ethnographic exhibitions," and they were a rousing success. Along with reindeer, he displayed round-headed Laplanders, barely four feet tall, in long skin coats and elfin fur caps. Then he dropped the animals and just showed the foreigners, starting with a family of Eskimos, who demonstrated their kayaking prowess; three young Somali men, who appeared in flowing gowns and what looks in one photograph like slender pine trees poking out of their heads; Indian dancing girls, Kalmucks, Singhalese, Patagonians, Hottentots . . .

Although the idea of sticking foreigners in a zoo for the amusement of the civilized world may seem a bit bizarre, it was picked up some years later, in 1906, by no less an organization than the New York Zoological Society, which exhibited an African Pygmy named Ota Benga from the Congo Free State. Twenty-three years old, he was four feet eleven inches tall and weighed 103 pounds. Then-president Hornaday always preferred to say the zoo merely "employed" him, but Ota Benga was displayed in a cage with the chimps, and an information label about him hung from the bars. *The New York Times* got hold of the story of a black man being exhibited at the zoo, however; a committee of black ministers in New York City protested; and Ota Benga was promptly taken off exhibit.

Out of the cage, however, the pygmy proved quite uncontrollable; at one point he grabbed a carving knife and ran around waving it menacingly in the park until he was forcibly disarmed. Later, when no one would give him a soda, he had a tantrum in which he started tearing off all his clothes. He was packed off to an orphanage, then to the Virginia Theological Seminary, of all places, and finally to a private home, where he found a revolver and killed himself with a bullet through the heart. He was thirty-three. One has to wonder if captive animals feel the same distress when uprooted from their native lands.

By 1888, Carl Hagenbeck's business had expanded to the point where he needed larger quarters than those available in Hamburg, and established a new park for his collection in Stellingen, a few miles away. His design for some of the buildings and the gatework displayed the exotic ornamentation developed at other zoos—the main gate, for instance, is a kind of sculptured menagerie, since it is decorated with stone elephant heads, polar bears, lions, and African warriors. But the interior was something quite new. Instead of indoor cages, Hagenbeck gave his antelope, lions, polar bears, elephants, and many other animals outdoor space in which to roam much as they would in the wild. And he went to great lengths to make the exhibits look natural, throwing up artificial mountains on wooden framework and, instead of using the standard iron bars, enclosing the territories inside watery moats. "I desired, above all things, to give the animals the maximum of liberty," he wrote in *Beasts and Men*. "I wished to exhibit them not as captives, confined within narrow spaces, and looked at between bars, but as free to wander from place to place within as large limits as possible."

Perhaps what was most revolutionary about his ideas was simply the belief that the animals could survive the harsh German temperatures. He had first caught on to this notion on a trip to England, where he saw a chimpanzee playing in the snow on the roof of a large tent. And at another menagerie, he saw that the monkeys were allowed out in the winter, but their quarters had an arrangement of flaps that allowed them to come indoors whenever they pleased. He noticed that the monkeys would go out even when the temperature dropped below zero. And he himself had accidentally left an Indian sarus crane out overnight in subfreezing temperatures. The bird was in perfectly good health in the morning. Hagenbeck was convinced that if the animals were allowed free access to the outdoors, they would not only survive, but they would be healthier than when perpetually cooped up in the tight and stuffy cages indoors. He had observed that such confinement often led to "mental depression" in his animals. He was convinced that most animals were naturally able to adapt to wide extremes of temperature, since plains animals, at least, often encountered a great temperature swing between day and night.

So he let many of his creatures go outside all winter long. *Beasts and Men* includes photographs of lions and Dorcas gazelles bounding about in the snow. There is one picture of an elephant so bundled up in animal skins that little more than its trunk protrudes, with the caption 60 BELOW ZERO. As close as Hagenbeck could determine, none of the animals died of exposure. His Indian panther liked the cold so much, it never went indoors but stayed out all winter long in the branch of a tree. A pair of lions had become droopy and lethargic indoors, but recovered dramatically as soon as they were allowed out. Large birds like the sarus and crowned cranes, various pheasants, and Australian black swans were all left out to good effect and new luster.

And just as significant, the animals were allowed to gather in natural groups, rather than separated in private cages. To judge by the dramatic breeding increases, this paid off in significant psychological benefits as well.

Other zoos took note. Inspired by Hagenbeck, the London Zoo constructed a panorama of reinforced concrete called the Mappin Terraces, and it created a separate free-roaming breeding facility in the country at Whipsnade. The Parque Zoölogique de Paris, near Vincennes, went barless in 1934. In the United States, the Milwaukee Zoo adopted Hagenbeck's ideas when it was built in the 1950s; today no zoo in America or abroad is immune to his influence.

The primate center was built in the Hagenbeck tradition. Its design is anti-architecture. Most of it, of course, is not a building but open land. As for the building itself—that is hardly a building either. Bradberry and his colleagues have done everything possible to make it disappear. Viewed from the walkway, as the elephants might see it, the building presents a long flat rectangle of reddish blocks. Although the façade seems flat as a billboard from end to end, it is in fact almost imperceptibly curved—an arc whose radius is somewhere in downtown Philadelphia—to give the building a feeling of tautness. The back side swoops in and out in two bulging and irregular curves, representing, perhaps, the wildness of nature. Seen from the top, as in the architect's renderings, the center has the exact outline of a pair of lips, as though the building took shape when an architect planted a kiss on the blueprint.

Bradberry and his associates shoved the structure a good four feet into the ground so that it wouldn't loom as a monstrous backdrop to the animals as they were viewed from the far side of the exhibit. And the building tapers to a sharp point at either end, as though it has been pinched off between the architect's fingers. In the process, the building loses a dimension. Instead of the usual four sides, there are only two. And, all the more remarkable, the façade runs 217 feet without a single window. Skylights admit natural light. The expanse is broken up and reduced to a human scale by a simple checkerboard pattern that runs along its length to a height of six feet. "The decoration scales down the whole thing and makes it go away," said Bradberry, adding, "although it really doesn't." Finally, the architects punched two big holes through the building for breezeways that lead the visitors' eyes, and bodies, to the animals. "It's sort of like the Champs Élysées years back," said Bradberry. "I've seen pictures before there was development beyond the Arc de Triomphe. You looked through the arch and there was just sylvan landscape beyond. You're just drawn to it. That is what is happening here."

* * *

If the outside of the building was reduced to the bare minimum, then the interior was stuffed with the furniture and equipment necessary to maintain a troop of about three dozen animals from eight primate species after hours. It was like a spaceship, so many things had to be crammed in. To start with, there was the problem of the primates' body odor. Their natural aroma taxes any conventional heating-ventilation-air-conditioning (HVAC) system. A standard residential house has one and a half air changes an hour. The primates required twelve. Yet, because the animals couldn't exactly put on a sweater when they got too cold, the HVAC system had to be carefully modulated so as not to put a draft on the animals. Also as on a spaceship, there had to be a backup HVAC and a backup generator, because if there were a breakdown, the animals couldn't very well run next door. Further, the rooms had to be designed not in the draftsman's standard two dimensions, but in three, since the animals climbed so freely on ropes and ladders. And to get the animals from their rooms to the main hatchway to the outdoors, the designers had to channel them through tunnels, which doubled as squeeze cages for veterinary exams. Each tunnel took up a tremendous quantity of space. "Do you put it up high so the keepers can walk under it, or do you put it down at the ground so you can always work at the animals' level?" Bradberry asked. "We decided to put it up halfway, so the keepers could work with the animals inside but still slip under it to get past, but that was a pain because the keepers are sure to bump their heads." Finally, what material for the primate beds? Cement is durable but cold. Wood is warm but the gorillas and orangs would tear it apart. Bradberry settled on chain-link fences coated with rubber PVC that could be slung between bars like hammocks.

Outside, more problems loomed. In keeping with the rest of the zoo, the designers wanted to maintain the look of an English garden. It also had to be escape-proof, indestructible, and nontoxic. "Every plant we picked for the garden, and we picked dozens, had to pass several tests. We asked, one, is it poison? Two, do the animals like to eat it? You don't want to give them anything they like to eat, because they'll just eat it up. And three, is it invasive? That is, does it grow like mad, so that every time they beat it down it comes back up?

"We wanted to put in the largest trees we could manage, eight to ten inches around. With maples, the animals will girdle them, strip the bark all around. They will do territorial markings and really mess them up. Sometimes an orang is just having a great time and he'll get up in the branches, go out too far and snap one off. So it gets wild."

In conjunction with the landscape architect, Hanna-Olin, Ltd., the designers ended up with honey locusts; oaks, which have the branch structure and form of some trees in the African rain forest; the weed tree ailanthus; ash; sycamore; the American hop hornbeam; Royal Paulownia, or empress trees, large and exotic with vast leaves; and devil's-walking-stick, which has held up well in the San Francisco Zoo, probably because it is so prickly.

To present a naturalistic shoreline along the various moats' edges, Bradberry didn't want to use concrete, but he wasn't sure what he should use because no other zoos had ever tried anything else. The firm settled on long sheets of black rubber that are used to contain toxic wastes. In case the water didn't restrain the animals, the keepers insisted on running lines of hot wires—akin to a rancher's electric fence but lower to the ground—along the shore. However, no one knew if that would work. "I can see it now," said Bradberry. "It's opening day at the primate center, and everybody's there, and it's *Whoa! There they go!* That would be something. And you know what? It could happen."

——2——

At the Rare Mammal House, gorilla keeper Bob Berghaier was pacing back and forth in front of John, the massive lowland gorilla. A little over thirty years old, Bob is himself fairly large, with a thick black beard tending to gray. Yet, as he crossed before John's gaze, his head was bowed, his face turned away, and his shoulders hunched in the obsequious manner of a charwoman passing before a king. John is in fact as royal a gorilla as one can imagine. He is big and powerful, with a Mr. Universe body (even Dian Fossey, no fan of gorillas in captivity, gasped with admiration when she saw him) and an imperious air. Now he followed Berghaier's movements back and forth like a spectator at a rather slow tennis match. The corners of his lips betrayed a feeling of mild contempt, but no more. After a few minutes of this, Berghaier slipped around to the back of the cage and gave John a cookie, which John ate glumly.

I asked Bob what he was doing.

"Just a little mitigation conditioning," Berghaier explained cheerfully. "I'm trying to get John to stop glaring at me every time I walk by. He tightens his lips and gives me a hard stare." He demonstrated by drawing his lips into a thin line. It made him look as though he were going to hit somebody. "Sometimes John slams his whole body against the glass. It's all a threat display. He doesn't really mean anything by it, but it makes a big noise and scares everybody to death. So I'm trying to get him to adapt his behavior."

It was a test of wills, really. John was trying to intimidate Bob, and Bob was trying to pacify John. Apparently, John didn't glare at everybody, just large men. Bob had thought for a time that his beard had something to do with it, so he shaved it off. John still glared at him, so he grew it back. It now appeared that John's feelings were more primal. He was the zoo's only grown male gorilla, and he was the alpha male for the zoo's troop of six—two adult females, whom he had taken for wives, and three offspring from the two relationships, plus John himself. As close as Bob could figure it, John saw Bob as a threat to his dominance of the gorilla troop. By his hate stares and threatening gestures, he was trying to scare Bob away, much as he would try to discourage any unattached males from wandering into his territory in the wild. Bob preferred to hang on to his job, but he was fully willing to submit to John's lordship, so long as John knocked off the tough-guy stuff.

Bob had just come back from vacation, and John had been unusually hostile toward him upon his return. At one point he had reached out through the bars to grab for him with his powerful fingers. "I think John thought he had driven me off," said Bob. "So when I came back he wanted to teach me a lesson." Bob managed to elude the big gorilla's grasp. He'd been grabbed once before, and he was not eager to repeat the experience. That time, John had hooked a finger around a belt loop and, with just that one finger, spun Bob around and yanked the keeper toward him. Fortunately, Bob was able to extricate himself before John got his other hand on him. Still, his heart was pounding.

I could see why John might not want Bob around. He had a pretty good setup, a small harem really: He had two females, Snickers and Samantha, and three young—one by Snick, and two by Samantha. Although all the adults were in fact the same age, Snickers was so hobbled by arthritis that she looked more like John's grandmother than his mate. You could practically hear her bones creak as she eased her body around the cage, and she had a real pot on her from lack of exercise. Samantha was in better shape, but she was hardly svelte; she retained the meatiness of a Russian matron. The two looked like a couple of bowling balls, with the young fry, Jessica, Chaka, and the baby Anaka, as billiard balls rolling all around them. Flat-nosed and round-eyed, Jessica was, at five, the oldest; and she was an imp, always testing the limits of her parents' patience. She often had to be brought back into line with a big slap of the hand from John. Chaka was tempted by Jessica's wayward behavior, but he was still too young to be daring. And Anaka was just a cuddly little baby; he hung on to his mother for safety like a castaway to a life raft.

The doors between the three end cages were opened up so that the whole troop could play together, but John generally kept to himself, sitting back on his haunches in the center of his cage, or loping about picking grapes, puffed rice, and seeds out of the hay. He did play with his children, but only in the tunnels "backstage," as it were, out of sight of visitors but in view of the keepers. Apparently, he had an image to protect when he was on public display.

Determined as John was, he had met his match in Bob Berghaier. One of the new breed of keepers, Berghaier had a dedication to his animal charges that was probably unmatched in the zoo. It had broken up his marriage. His wife wanted him home, but Bob felt his first loyalty was to his animals. He worked long hours with the creatures, and when he was home, he often had his nose in a book or journal about them. His wife never understood that. In the four years of their marriage, she came to the zoo only once. So she left him, and he wasn't sorry to see her go. "A job like this can really take over your life," he explained. "My wife and I didn't see eye to eye with it. You know, this job really becomes your identity. I'm not Bob Berghaier. I'm a zookeeper—even to my family. The first thing my parents ask me is, 'How's the zoo coming?' I can really monopolize conversation."

Liking animals from a young age (although he was terrified of cows), Berghaier had enrolled in a wildlife-management program at Penn State, then added a night course in biology at St. Joseph's University. His plan was to go into field conservation, but he realized that he enjoyed contact with people, too, and at twenty-two he came to the Philadelphia Zoo as a keeper. He still kept up with his academics, though, and subscribed to a variety of esoteric publications like the *East African Journal of Ecology*. Of all the zookeepers, Bob was probably the most thoughtful, almost scholarly. Other keepers had taken to calling him the "Professor." Unlike the others, he kept pretty much to himself, just as John did; he devoted his few spare moments to devising different schemes to improve the lot of the animals under his care. For instance, he was the one to suggest putting hay and cardboard boxes into the gorillas' cages for them to play with. Other keepers had opposed it, because it meant more work, but Bob had finally prevailed.

Seeing Berghaier's dedication, the zoo had tried him out with the elephants. That assignment had given him a lot to think about, particularly after a run-in with Kutenga. "She threw me into the wall, and she was ready to kneel on me when they got her away," he said. "It was potentially fatal. It rearranges your mind to realize at such a young age that you're not immortal, I'll tell you that. With the elephants, you have to fit into their herd structure and try to become the dominant animal without the ability to discipline the other animals that the real dominant males have—by busting heads. The keepers that make it are the wilder ones. You can see it in Jimmy McNellis and Dave Wood. They are bull elephants—ever noticed? When you push them, they'll take it for a little while, but then you see the bull elephant in them rear up. You can almost see the trunk unfold and the ears start to flap. They're wild."

And Gene Pfeffer? "He can be like that, sure." Would he make it? "That's hard to say. Gene really wants to do it. He's incredibly obstinate. When you almost get killed by an elephant, it does a number on your head. The fact that he got clocked pretty good and he still wants to get back in there with them—that tells me something."

I asked him what it took to work with the animals.

He wrinkled his forehead as he pondered the question. "I think it's a spiritual quality," he said. "It's an in-tuneness with the animal. I don't really have it. About five years ago I realized that all I had was book knowledge. I didn't have any of the real kind. I've been working on that ever since, trying to increase my perceptiveness and improve my technique. See, some guys are just naturals. Dave Wood—he's one—and Ray

Hance over in African Plains is another." Ray was the one who had confronted me on the question of the Cat Man's sanity; I had also marveled at his athletic grace in working the springbok. "A lot of it is simply the way you move, and Ray naturally moves well. He's never jerky or quick. There's more flow to it, fluidity. He seems to be in harmony with the animal. I suppose it's a kind of empathy, a special feeling for the animals. You go with them, not against them. It can be hard otherwise. I think one of Chuck Ripka's problems with the elephants is that he is too herky-jerky. You can never be obtrusive, particularly here with the primates because they are so sensitive."

Despite his interest in progress, Bob was leary of the new primate center. "I have to admit I'm disappointed," he said. "It's going to be a lot smaller than I'd thought." When you added up the square feet, the cages in the new indoor space would be smaller than the present ones here in the Rare Mammal House. And even the outdoor area looked more cramped than he had imagined from the drawings. "I guess there just wasn't a lot of room there," he said wistfully. And he was worried about the moats. Gorillas don't swim, he observed. In the wild, rivers form natural limits to their territory. He was afraid that a gorilla might get frightened, fall into the water, and drown before it could be fished out. "I think they made the moats too deep," he said. He'd been the one to ask Jim Bradberry to install the hot wires to keep the animals up on their islands.

Roseann Giambro, the head primate keeper, was even more upset. "When you get right down to it," she said, "it's still the same old bathroom tiles on the inside." Jim Bradberry had weighed all the alternatives, but decided that tiles were best for durability and hygiene. And he had selected a sky-blue color because something light would make the animals show up better: Light yellow would have made them look jaundiced, light red—or pink—would have been too trendy, green too outdoorsy, and there was nothing left. Roseann wouldn't hear of it. "You'd think that for all that money they could have done better," she said. "If you ask me, I think they should leave the primates here at the Rare Mammal House."

Bradberry was frustrated to hear these complaints. "Roseann came in out of left field," he said. "She started raising objections in the fall, and I'd never seen her before. She said she didn't like the blue. 'Well,' I told her, 'it's not like shopping for clothes. We had a meeting about that three weeks ago, and where were you?' To be honest, when we got into the project, I would have complained about the tile, too, but the fact is there is no other choice."

Part of the problem, Jim explained, was that the keepers now on the job were not the ones who had been supervising the design process all along. One of them, Rick Beyer, was dead. Others had moved on to other jobs. The current crop was probably more conscientious than the previous crew, but it was too late to do much about their objections. Some of the complaints, Jim admitted, were legitimate, such as the location of an electrical junction box inside one of the gorilla bedrooms, and the presence of a couple of blind corners. But others, he thought, were simply hysterical, like the idea that the new primate center doors were smaller than the old ones at the Rare Mammal House. He measured the new ones, and they were larger by a couple of inches. As for the size of the bedrooms, he realized that they weren't the biggest bedrooms, but that the animals would only be in them at night. And he agreed that the walls were "not the most attractive," but they required the least maintenance. "It was all a trade-off," he said, "and we did as well as we could with the money available."

Roseann was not appeased. She felt her gorillas had been shortchanged, and she brooded about it increasingly as the World of Primates was rushed toward completion.

3

Spring came to the zoo the last week of March, when the temperatures soared into the seventies. The air lightened, the breeze, usually laden with animal odors, carried a perfumed scent, and the garden was filled with birdsong. The crocuses gave way to daffodils, forsythia, and flowering dogwoods; and hundreds of fruit trees—Japanese cherry, crabapple, even pear—burst forth with luscious blossoms. Bird Lake looked especially gay with a garland of apple and cherry blossoms circling its shores. In the nature walk of Penn's Woods, the columbine, ironweed, wild geraniums, and other traditional Pennsylvania wildflowers were out in profusion, causing the hummingbirds to make a detour on their way from South America to Nova Scotia.

Robins in the branches of the oaks and sycamores were no longer a sign of spring; since they had discovered such a plentiful food source at the zoo, they stayed through the winter. But the warblers still vacationed in the South at wintertime, and they returned in force, along with goldfinches, grosbeaks, sparrows, and an occasional bluebird, and, despite the zoo's chemical intervention, more pigeons than one would care to count. And the peacocks and peahens were at large, much to horticulturist Chuck Rogers's consternation, after taking shel-

ter behind the Carnivora House for the winter. Everywhere
the zoo was bright and glittery and festive.

With the warm weather, the visitors came back. All the
zoo staffers claim they enjoy winter because they can get so
much done. The vets can drive along the sidewalks during the
day without fear of running someone down, and the keepers
don't have to post themselves by their enclosures' fences to
remind visitors not to feed the animals (since many are on special diets, and all need to have their daily intake carefully monitored). And yet without the swarms of visitors, the zoo had
seemed a little drab and lonesome. It was uplifting to see the
humans return—the mothers wheeling their newborns, the
teachers herding their school groups, the young couples walking about arm in arm.

As for the animals, they were busy tidying themselves up
for the new season. They had turned the whole zoo into a
barber shop. Great tufts of their winter coats wafted about in
the air. Clumps of white and gray and brown skittered across
the enclosures and out onto the walkways. Normally neat and
well groomed, the animals seemed to be parading in their un-
derwear. Poor Matilda the camel was the worst. With the coming of spring, the keepers had finally released her from her
stall. And now she suffered yet another indignity, losing her
hair—a clump here, a clump there, as though she had gone to
some hack barber.

The antlered animals started dropping their antlers. Both
the web-footed barasingha deer and the little muntjacs lost them.
Horned animals—antelope, tahrs, giraffes—hang on to their
horns. But the antlers sometimes toppled off one on at a time,
leaving the animals drastically lopsided. Within a few days, new
antlers budded atop their heads; over the summer, they would
sprout and fork like the branches of a tree.

As the zoo warmed up, the zest returned to the animals
after the long winter doldrums. The ducks, geese, and swans,
huddled by the shore for the most part over the winter, resumed paddling around Bird Lake just as the boatmen of the
Schuylkill navy returned to their river across the expressway.
Mopey-Dopey and his friends were out in their pen, and the
great tortoise was starting to feel those familiar sexual pangs.
The zebras shifted nervously about in their stall down at Afri-

can Plains; and the antelope resumed their graceful frolicking
as though they had never left it. Even the elephants seemed
peppy now. Only the polar bears and the wolves were slowed
by the new developments. Polar bears never shed, and the
wolves were laggard about it. As I watched one sunny afternoon, the four wolves lazed about down in Wolf Woods as
though they had been drugged. Trillion, however, had inched
a little closer to Taboo.

Early spring is also the time when zoos ship animals in
and out. It is too hot to move them in summer, too cold in
winter, and the fall is less appealing because the zoo doesn't
get a return on its investment until the new visitor season. So
the curators move them in the spring, and it seemed that every
day some animal was coming or going.

The incoming animals didn't always act pleased to be here,
however. One day in late March a zebra was trucked in from
a zoo in Jacksonville, Florida, to replace one that had died over
the winter, and the zookeepers had never heard such a fuss.
The zebra was pregnant, too. Her name was Dotty. But then
all zebras are a little screwy—fierce, high-strung animals. Few
zookeepers like to tangle with them. Gunther Schultz, attired
like a cowboy in boots and jeans, left her off in his big modi-
fied moving van, a kind of ark on wheels. He had trucked in
Xavira years before. There were buffalo and blackbucks and
nine other species of hoofstock secreted in the van's various
compartments, all bound for different zoos up the East Coast.
The van was so big, Schultz hadn't been able to clear the gates
by the Penrose Lab for the turn onto the interior road that led
to African Plains. The truck got stuck halfway through. The
zoo had deployed a front-end loader to hoist a crate to the
zebra's door. The driver left it there, with the connecting doors
open, in hopes that Dotty would leave her van compartment
and come nosing into the crate out of curiosity. No chance.

"She wants the devil she knows, not the devil she doesn't
know," said Mike Barrie, who had emerged after performing
surgery on a woodchuck to watch the action. Dotty stayed put.
With the open mouth of the crate abutting the doorway to her
compartment, there was no way anyone could grab the zebra
to haul her in. Of course, nobody would have much wanted to

anyway. Zebras are about as calm as bucking broncos, and they harbor grudges. Besides, Dotty was so riled up from the trip, she would have bitten and kicked her grandmother. Gunther tried to encourage the zebra to move into the crate by poking her with a stick through a porthole in the far side, but Dotty merely bit at the stick and refused to budge. She kicked the interior wall of Schultz's rig a few times to register her displeasure. The noise attracted a number of spectators, all of whom started to make suggestions.

"We need a balloon," said the new curator of mammals, Karl Krantz. "Stick it in and blow it up. She'll come flying out of there."

"I think an umbrella would be better," said Mike. "Poke it through and then open it up."

Keith Hinshaw had been watching the spectacle with a sinking heart. He knew that once the animal was poked with a stick, there wasn't much hope of her trotting out on her own. "I think we'll have to use the anesthetic," he told Karl. Karl agreed. He went back into Penrose and returned with a pistol. It looked like an automatic, actually, and it came with a holster, but he didn't put it on. Still, Keith had a Gary Cooper air as he advanced on the van with the gun in his hands. Instead of a bullet, the gun bore a load of anesthetic M99. He assumed the FBI stance, holding the gun with both hands and his feet spread, as he drew a bead on the animal through the tiny porthole into the van compartment.

Then he fired. There was a silence for a few seconds. Then—ka-BAM! The zebra exploded in fury. Crash! Bang! Slam! Pow! It was like a cartoon. No one except Keith could see her, but Dotty was obviously going wild inside. To judge by the sounds, she was banging her hooves on the walls, floor, *and* ceiling.

Keith walked back toward us, his head hanging slightly.

"I guess she didn't like it," said Mike.

"Nope," said Keith.

Finally, the sounds quieted down, and in a stupor the zebra wandered into the crate and slumped down as the door crashed behind her.

The front-end loader carried the crate around to African Plains. Several keepers, including Dan Maloney and Gene

Pfeffer, helped shift Dotty onto a heavy canvas and then half carry and half drag her to an outdoor pen that had been cleared for the new zebra's arrival. There they left her stretched out on the ground. One leg poked out stiffly, and her lips were quivering.

Weakened as she was by the anesthetic, Dolly was still something to behold with her dazzling Op-Art black-and-white coat. Ethologists have been puzzling for years about the utility of the zebra's unusual markings. Do the stripes deter flies? Blind lions? Help regulate their internal temperature? The current theory is that the zebra's coat serves to enhance group cohesion. Zebras find the nearly electric pattern of black and white in the herd naturally enticing; they are drawn to it the way we are pulled to the flickering images of television. The striped zebras consequently band together; the occasional unstriped zebra, in fact, is shunted away to the outer fringes of the herd.

Karl Krantz looked over the merchandise. "She's a good-looking animal," he said. "Good thick mane, nice color. If she calms down, she'll be fine." He pointed to some scuff marks on her side where she had bumped up against the side of her compartment. "Once we get these smudges off her, she'll shine up really nice."

Mike Barrie prepared a needle to inject the antidote to the M99. "Everybody better clear out," he said.

"How fast does the antidote work?" I asked.

"Fast."

"Really fast?"

He looked straight at me. "Let me put it this way. With polar bears in the Arctic, the polar bears will beat you back to the helicopter."

I left the enclosure.

Mike gave the injection, and then he scooted out, too. We watched from the far side of the fence as the zebra climbed to her feet in seconds, shook herself, and then trotted around her pen to check the place out—the weathered boards that ringed it, the few tufts of beaten-down grass on the otherwise barren earth.

I took the opportunity to ask Gene about the elephants.

"Everything is coming along fine, just fine," he said in a way that closed off further discussion.

Keith stayed there, eyeing the zebra like a baseball scout sizing up a prospect. "She looks okay," I said.

"Yeah, but we're lucky. That's not the way it's supposed to be done. I didn't like the fact that they left her to bang around all that time this morning. You give her a quick try, and if that doesn't work, then you anesthetize them. They can hurt themselves pretty badly in those vans, and you want to get them out and into the enclosures as quickly as possible."

I walked back with some of the keepers, but Keith stayed behind by himself, watching Dotty to make sure she was all right.

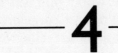

—4—

Karl Krantz was bent on breaking down the longstanding social distinction between curator and keeper. Theoretically, curators were in charge of a broad class of animals—the mammals, in Karl's case—and keepers looked after a small group of animals in their care. As with doctors and nurses, the two jobs might overlap, but the prestige levels were separate. In sympathy with keepers, Karl often wore blue jeans and a work shirt, instead of coat-and-tie curatorial clothes, and he didn't mind getting down and dirty with the animals. In general, he acted more like a ranch hand than a behind-the-desk professional.

Young and boyish with thick and curly dark hair, Karl had come to Philadelphia from the National Zoo in Washington, where he had been the curator in charge of education. He'd arrived during the winter, replacing Dietrich Schaaf. He was a little awed by his current responsibilities. At the National Zoo, he pointed out, there would be three curators doing his job: one for primates, one for carnivores, and one for small mammals. Here he was in charge of the whole thing, and he often found himself reaching for animal encyclopedias to get a grip on the less familiar species.

Right now, up in his Penrose office, he was particularly

anxious about the impending move to the new primate center. He'd been grinding his teeth again in his sleep, he told me. His TMJ syndrome, a stress-related disorder of the jaw, had been acting up, and he was going to talk to Dr. Tinkelman about it the next time he came in. He couldn't believe all the primates would make it across to the new exhibit safely.

One thing that did not worry him was the apparent marital problems of the Siberian tigers George and Martha. "They're still naïve animals," he explained. "They've still got to work things out. Contrary to what people think, you know, some practice is required for successful sex." Then he shouted toward the next room: "Isn't that right, John?"

Silence.

"I said, ISN'T THAT RIGHT, JOHN?"

"How's that?" replied a voice. It was John Groves, the curator of reptiles.

"Practice is required for sex."

"How would I know about that?" replied Groves. "You're the married man." Groves was divorced.

"You're the one with the kid."

"Well, then, yes, I guess you need some practice, sure."

"See?" Karl continued. "I really think that's our problem. They'll get it together. Also you have this basic problem of personality— Hey! Look at THAT!"

He was eyeing something over my shoulder, and I quickly swung my head around. It was an insectarium on a shelf by the window. Thinking of adding to the Small Mammal House a small exhibit of some emperor scorpions, Karl was raising a couple of the large, furry creatures, a metallic blue under the infrared light of the insectarium. He gazed down at one of the scorpions in wonder. "It's started to drink!"

Sure enough, one of the emperor scorpions had scuttled over to the water dish and started lapping at it. Apparently, neither creature had moved during the several weeks that Krantz had had them. He was afraid that they might not make the best exhibit.

I asked him how he had decided on them.

"Well, it was a gamble whether I should keep the car, or take what was behind the door, and the audience was screaming, 'The door! The door !' So I went for the door, and this—"

He didn't finish the story because he was so distracted by the hairy insects. He was peering into the cage, radiant as a kid with his first lightning bug in a peanut-butter jar. He gazed down at the scorpions adoringly as though he wanted to kiss them on their furry backsides. "I'm so happy," he repeated. "He's taking a drink!"

—5—

Karl Krantz was busy netting some otters for their annual dental checkup a few days later when Dr. Tinkelman came in. Karl took a moment to ask him about his TMJ. Tinkelman said there wasn't much to do about it except to relax. "Great," said Karl sarcastically. "How do I do that?" Tinkelman couldn't answer that one.

Located in front of the TreeHouse, the otter pool was round and deep; it had a rocky slab in the middle for the two otters to perch on and sun themselves. Workmen had drained it in advance, for netting an otter would be nearly impossible in the water, they are such quick and elusive swimmers. They darted about the pool like trout. Even without the water, the otters were still mighty shifty as they scampered about the floor of the empty pool on their little paws, and Karl ran madly along the tiled floor, with his net down low like a field-hockey player trying to catch them. From the way he guarded his ankles, it was clear that he was mindful of the otters' sharp teeth.

"Is this why they gave me insurance?" he yelled up to Mike who was watching him from a safe distance.

"They *gave* it to you?" Mike yelled back.

Finally, Karl and some keepers nabbed one of the otters while the other scrambled to temporary safety up onto some rocks. Mike Barrie injected the captive with anesthetic through the net. In a few moments, the body went as limp as a mop. Then some keepers carried it back to the surgery room, where Dr. Tinkelman set to work scraping off the dental plaque—it looked like the remains of a Sugar Daddy—and the vets went after the remaining otter down below.

"You don't worry about him waking up too soon, do you?" I asked Tinkelman.

"Oh no." He continued to pick at the plaque. "That never happens."

Mike and Keith were giving the otter a vaccination for canine and feline distemper, and for rabies. Then Mike took the opportunity to tattoo the otter's ISIS number inside its thigh with India ink. Eileen Gallagher clipped off some of the otter's fur to mount on slides for an educational display. Her face looked a little puffy, and she was sniffling a little, as though she had been crying. I was afraid she might have lost another animal.

As usual, the vets were chatting.

"You know that Prevost's squirrel we shipped down to Houston?" asked Mike as he worked the ink injector.

"The male?"

"Well, no. Turns out he was female. She gave birth down there."

"She *must* have been a female, then."

"Yeah, guess so."

Just then, the otter's jaws snapped shut, and Tinkelman jerked his hands back. "Hey! He's awake!"

Keith hurried over. "We should use the halophane," said Keith quickly, referring to an anesthetic gas.

Mike jumped to hold the otter down while Keith inserted a tube into the otter's throat to force some gas into his lungs. Eileen monitored the dials to regulate the flow. In a few seconds, the otter had gone limp once more. "Ah," said Tinkelman, "check that about animals never waking up too soon." He said it was better to use the gas anyway. It gave him better control of the animal's level of consciousness, and it wore off more quickly after the operation than fluid injections. As a precaution, he inserted a pair of springs to hold the otter's jaws open. A few minutes later, the animal went a little *too* still. "Wait a second," said Tinkelman. "This guy's stopped breathing!"

Eileen adjusted the dial a bit, and the otter drew breath once more. "I guess we went too deep," he said. Then he went back to scraping.

6

Afterward, Eileen was puttering around the lab, still sniffling. I asked her if she was all right. "Oh, it's these allergies," she said.

"Allergies?" I asked. I hadn't known about them.

"Yeah, I've had them for a while, but they're acting up. It started with the dik-dik that I took care of over the winter, the one that died."

Eileen had taken care of it for a few weeks, but the dik-dik—a small, delicate antelope—had died one night at her apartment. She hadn't been able to bring herself to stash the corpse in her refrigerator, so she had put it outside in the cold, then brought it in for Dr. Snyder to do a necropsy in the morning. She had been so preoccupied with the dik-dik's problems, she hadn't paid much attention to the tiny red bumps on her forearm.

The vets told her to see an allergist. "I went in wearing my zoo clothes," Eileen said, "and everybody in the waiting room thought that was really funny. A zoo worker with allergies, you know, hah, hah, hah. When I got in to see him, the doctor couldn't figure out what was wrong, because he could only test for an allergic reaction to dogs, cats, and birds. You know, common animals. And I'm fine with Roxanne, my dog. They don't know about exotics."

Her latest outbreak had been caused by three new baby binturongs she was taking care of. She beamed as she told me the news. "People have asked me if I have a new boyfriend," she said. "And I say, no, I've got some binturongs."

It had been a little awkward to get them, though. By rights, Ann Hess should be the one to take charge of the surrogate mothering. But ever since she'd had to send off her beloved Ralph to the Bronx, Eileen had been telling the vets that if there were ever any more binturongs to raise, she wanted to do it. Beauty, the female binturong in the Carnivora House, had become pregnant again over the winter. Since Beauty hadn't attended to her earlier offspring, Ralph and Alice, Eileen assumed that these babies were going to be hand-raised from birth, too. She could hardly wait. "I was so excited," she said. "Everyone at the zoo was coming up to me and calling me Mom."

A few days before the birth, however, Keith called Eileen and Ann Hess into his office and told them that raising the binturongs was too big a job, and he had decided to divide it between them: Ann would have the animals in the nursery during the day, and Eileen would have them only at night. Eileen, who had already been disappointed that Ann got to raise so many more of the zoo animals, couldn't believe what she was hearing.

"I was hysterical," she said. "When I get that way I can't talk, I just babble. So I babbled. Then I started throwing up, I was so upset."

A few days later, the babies came: three kittenlike binturongs named Al, after Al Porta; Abe, because they were born on Lincoln's birthday; and Ashley, because they needed another name that began with A. Together they were known around the zoo as the A-team. They were left with their natural mother that first day in hopes that she might reform and nurture them herself. The next day, Al Porta unknowingly came into the Penrose Lab with a flower for Eileen as the new binturong mom. Eileen burst into tears. "But they took my babies from me!" she sobbed.

Eventually, she was able to talk to Keith about her feelings. "I'm very, very hurt," she told him. "I can't get over this. It's very tough. It's like you gave my kids away! I can't be a part-time mother." Keith, analytical as ever, replied that it was Ann's job to do the mothering, and that he couldn't take the binturongs away from her now. "Her feelings would be hurt," he said.

"What about *my* feelings?" Eileen replied.

Just then Wilbur Amand came in and, not knowing what the two were discussing, asked Eileen, "How are your babies?"

"I just don't know!" she yelled and stormed out of the room, leaving Wilbur to wonder what he'd said wrong.

She took a two-week "mental health" vacation. She visited a friend who had in fact written a thesis about binturongs. "He knows how attached you can get," she said. The two of them spent much of their time looking at photographs of binturongs.

Ann Hess raised them while Eileen was away. Ann drove me to her home in New Jersey to see them one night. She never mentioned any of the trouble that she was having with Eileen.

Her house was a small metal-sided building in a little housing development just off the highway. She lived alone with her Rottweiler, Jenn, four stray cats that she had taken in, and, currently, the three baby binturongs. The dog took to the binturongs very well. "She's used to them by now," said Ann. But the cats were a little suspicious. Ann had brought the binturongs home with her from the zoo in a plastic traveling case, which she set down in the kitchen. The cats had sniffed it with interest, but as the three little binturongs nosed out, the cats backed off, startled. The binturongs looked like kittens, just as Ralph had. They had dark, fluffy fur, coal-black eyes, black button noses, and long, ratty tails that twisted around like treble clefs. All three were virtually identical. However, Ann could tell them apart because Al was a little grayer than the others, and Ashley had a smudge of white on her face. Ashley was also a ne'er-do-well; that night she immediately peed on the linoleum floor as soon as she emerged from her traveling crate.

"You were supposed to have went in there," Ann scolded, referring to the paper-lined cage, as she mopped up the puddle with a paper towel.

Ann showed me around the house, pointing out the spare bedroom where she had raised Aava the aardvark, Kiki the gorilla, an orang, some racoonlike kinkajous, and all the rest, and gesturing to the backyard where she had let her grizzly bearcub lumber and her kangaroo hop. She had become quite a conversation piece to her neighbors, and she often found children hanging by the fence to see what she'd brought home.

The binturongs were skittering across the floor when we returned. "They're a lot to deal with," she said. "Especially in the morning. I'm not so cheerful then." Ann prepared the binturongs' dinner of what looked like baby food. She set the meal out in bowls, but had to scoop Ashley up to feed her by hand. "She's not real good at eating on her own yet," she explained. When the others were done, they gave out soft *wff* noises as they sniffed about. They tussled with a big rubber pretzel, skated around the slippery linoleum some more, flopped into each other's food bowls, and then took to jumping off their hind legs into the air as though they were pouncing on some invisible prey. Ashley seemed to be leading the others in these imaginary muggings. "It's play attack," Ann explained to me. Then she plucked Ashley into the air and looked her full in the face to make sure it really was Ashley. "You think you're so tough," she said playfully. "You think you're so tough."

Al meanwhile wandered over to investigate my shoes, and grabbed onto my pant leg.

Now it was potty time, and Ann rubbed each of the animal's rear ends with her finger through a cloth to get them to "go." It was as easy as squeezing toothpaste from a tube; they went, one after another.

"It's a lot easier with these smaller animals," said Ann. "I had to keep the drills and gorillas in diapers. Otherwise it can get messy. They go, and then they walk in it, then they walk on you."

Next, she prepared their bottles of formula. Ashley came over and cuddled with her for a moment. "Ashley, what? What?" she said in a mother's singsong. "Oh, you want this?" She produced the bottle. "Here you go. Here." Ashley was curled up in the crook of Ann's arm. Her forepaws grasped the bottle. She sucked hard on the nipple. "There," said Ann. "That's better."

When Ashley had had enough, she came over to me and leaped into my lap. In a trice, she was up my jacket to my neck, where she started nuzzling my ear. Her whiskers on my earlobes sent chills up and down my body. She tried to ascend to the top of my head, but couldn't manage it. Finally, Ann rescued me. "Ashley," she said, "it's not nice to climb on our guests." She set her back on the floor.

It was almost the binturongs' bedtime. As she drove me back to town, she told me about her index finger, which had been bitten by the woodchuck over a year before. I couldn't help noticing the finger; it was red and stuck out stiffly from the others as she clung to the steering wheel. She couldn't bend it anymore, and she was always catching it on things. Recently, the doctor had told her she would have to have it amputated. "And all because of a woodchuck, too," she said unhappily.

When Eileen returned from vacation a few days later, she had resolved to let Ann raise the binturongs by herself. She couldn't be a part-time mom. Then she found out about Ann's impending operation, and agreed to take the animals until Ann recovered.

So, for the moment she had her binturongs, and they made her very happy. It was like old times with Ralph in her apartment. "I love them so much!" she said. "I'm like a ten-year-old." The animals were unbelievably springy. "When I open the refrigerator door," she said, "they're on the second shelf." And they seemed to respond to her. "As soon as they hear my voice, they start crying for me."

Recently, however, she had started discovering the red welts sprinkled across her body, and realized she was allergic to the animals she loved most. "They wrap their tails around my neck. Then in the morning, I find these terrible red marks, like a rope burn there. And I get little scratches every time I pick them up. The worst thing is that I like to kiss them. But it makes my lips burn and turn red. And I think, 'O my God! I can't even kiss them!' So now I have to bundle up when I'm with them. I wear my long blue scarf when they climb on me, and I can't cuddle as much. And I have a lot of Benadryl on hand. But that's okay. I figure that's just the sacrifice you make. What can I do? I'm addicted."

"You didn't have this problem with Ralph?" I asked.

"No. Some allergies you don't get until you're exposed for the second time. Or maybe Ralph was just different."

But Eileen knew that as soon as Ann was better, she would have to return the binturongs to her. She still couldn't share them, so it would be better this way. She was trying to get used to the idea.

"They aren't like Ralph," she said. "They're not as quick. And because there are three of them, they pay more attention to each other than they do to me. Ralph was more dependent on people because it was just him." She weakened again. "But they are cute. And they do many of the things that Ralph used to do like climb all over the place and nibble the corner of my eye." Then she firmed up once more. "But they're not Ralph; that's what makes it better."

7

Now that the weather was warming up, I strolled down to the Children's Zoo to see how Sue Jendrowski Smith was coming along with her trained sea lions. The crowds would be coming in for the shows soon.

Generally, it was considered a poor idea to train animals to perform. The shows smacked too much of chimpanzee tea parties and trained bears and other indignities. But that left aside the fact that many of the animals did seem to enjoy the training and the public loved the shows.

The performances persisted elsewhere. At the Portland Zoo, an animal psychologist named Hal Markowitz had gone so far as to teach Diana monkeys to earn tokens by pulling a lever, then to buy their dinner by pressing the tokens into a slot; to get polar bears to order their dinner by growling for it; to teach elephants the difference between light and dark; and to get a mandrill to compete with human visitors in a speed reflex game involving a lighted display (the mandrill did pretty well).

Nevertheless, the Philadelphia Zoo had rejected such behavior training as unnatural and unfairly stressful to the animals. Wilbur Amand dismissed the motion rather brusquely when I raised it one time. But some others reminded him that he was touchy about it only because he had once lost at tic-tac-toe against a chicken. "But he was a very good chicken!" said Wilbur.

Still, the zoo did put on sea lion shows every day. Sue Smith was the Philadelphia Zoo's answer to Joan Embery, the San Diego Zoo's well-known animal trainer, performer, ambassador, and frequent guest on *The Tonight Show*. Sue had been a veterinary technician like Eileen Gallagher, but when she started doing the animal performances, she developed a marked public persona. Even in private conversation, she spoke a little louder than necessary and *landed* on certain *words* as a lecturer might. And when she walked, she tilted her head back and carried herself lightly on the balls of her feet. The zoo's border collie, Lassa, trained to herd the Children's Zoo's collection of pygmy sheep, invariably followed at her heels. "I have a good relationship with her," Sue told me in the Children's Zoo office one morning as Lassa looked up at her. "I can tell because whenever I'm in, she's all over me." As if on cue, Lassa jumped up on her and licked her face.

Today Sue was excited because she had finally managed to get Molly to jump through a hoop. Molly was the Philadelphia Zoo's veteran sea lion. Sea lions are distinguished from the better-known seals by their external ears and their ability to flop their rear flippers forward. Sea lions can walk (although not too well) on land, while seals lie on terra firma pretty much like sausages. Sea lions even look a little like lions, if you can imagine lions that have slipped into wet suits and flippers for a swim. They are, in fact, carnivores that have taken to the sea, and in their vestigial limbs, now flippers, you can still see the last traces of their landlubbing way of life.

The zoo had owned Molly for several years, but no one succeeded in training her until it acquired Salty, a male, from the New York Aquarium. Salty had needed a change of scene: He had developed stress ulcers in New York from being bullied by a larger sea lion named Jaws. "He liked it here from the first day," said Sue. He was especially pleased that the zoo continued to indulge his taste for expensive mackerel. He had developed a passion for the fish in New York when they were the only things he could eat. He earned his keep, though, for he proved to be an old pro at performances, and he had inspired Molly. Previously, she hadn't done much more than wave, a trick picked up from Salty's predecessor, Chester. But under Salty's influence she had learned to slide along the pier, salute, clap for herself, give Sue a kiss, catch a Wiffle ball on her snout, and now she could jump through a hoop four or five feet in the air.

"To teach them, you work with their natural behavior," said Sue. "You do it by approximating," she said. "Like with getting her to jump through the hoop; first I got her to touch it with her nose on land, then to touch it in the water, then go through it to her shoulder, then to her flipper, then *all* the way through it underwater, then I lifted the hoop up higher and higher, and finally, last week, she jumped through it in the air. But it took a long time, about a year."

Molly continued to be standoffish in some respects, however. Sue had learned to let Molly touch her if she wanted to, but to hold off reaching out to Molly herself. "Touch shows trust," she said. She hadn't entirely earned Molly's yet.

Sue had learned the art of training sea lions in a three-and-a-half-week course she took at the New York Aquarium the previous year. And she was the one to develop the zoo's show. Previously the zoo had contented itself with an act involving a trained mule named Dixie that had previously been employed in a circus-clown act. Dixie could walk backward and jump into the air. But that act paled beside the sea lions, and Dixie was put out to pasture at the Children's Zoo. She had put on weight since then, and developed a hay belly.

Sue looked at her watch—it was showtime! She charged out of her office, with Lassa and me trailing after her, pulled on her waders, and took up her position on a wooden pier by the large kidney-shaped pool. The pool was once the site of a miniature ark, with a dozen animals crammed into its hold, but the zoo did away with the exhibit the previous year to make room for Salty and Molly's show. Just a few visitors were around, but it was early in the season yet. The sea lions were dashing around the pool like underwater missiles, but they snapped to when they saw Sue on the platform.

She introduced herself and the sea lions to the crowd. "Okay," she said, "Salty, now wave." Salty rolled over onto his side to wave. She blew on her whistle, said "Goooood," and threw him some fish, which he wolfed down. Molly did the same, and got the same treatment. They acted like kids eager to show off for grown-ups. Then they went through their routine: Salty caught some rings around his neck, balanced a basketball on his nose, jumped through a hoop (but not very high, that's one of his problems), and barked out a scratchy "Hello" on command. Molly did some of the same tricks, with a little less assurance. Then the show was over.

"Give yourself a hand, Salty," said Sue.

Salty hopped up on the platform, arched his back, and slapped his rear flippers together. The crowd burst into applause.

"Give yourself a hand, Molly." Molly looked at her and sped away. *"Molly."* A little impatience crept into her voice. "Give yourself a hand." Now Salty hopped back into the water, reemerged, and clapped. No reward. "Give yourself a hand, Molly." Finally, Molly flopped up on the pier and clapped. "Goooood," said Sue, and gave her some fish as applause rose all around. The show over, Sue passed her whistle and fish pail to an assistant and walked off. The sea lions watched her go for a brief moment before turning their attention to the fish pail, not realizing it was empty.

——8——

As the big moving day approached, the gorillas were carrying on as always in the sixties-style Rare Mammal House, completely ignorant of the coming cataclysm. John haughtily knuckle-walked about his cage, picking out puffed rice from the straw, Snickers cuddled her baby, Anaka, Jessie acted up like a rambunctious teenager, and Samantha strained to keep up with her gorilla toddler, Chaka.

With many zoo animals, you have to watch carefully—and patiently—to catch the unfolding melodrama. With the gorillas, it is soap opera, broadly played. Gorillas have a gift for theater rivaled only by their noisy neighbors, the chimpanzees. They show love, tenderness, anger, pride, and fear in unmistakable ways. And they are so lavish with their emotions that even the casual visitor is quickly absorbed by the tale. While the fundamental order of the troop, from the fearsome big daddy, John, down to his fluffy youngest, Anaka, remained stable, there were enough conflicts within it to fuel many a *Dallas* episode.

Recently, for instance, Snickers had it in for Samantha. It was hard to know exactly why. She might well have been jealous. Samantha was the younger and prettier of the two grown females, and she had, enviably, two young sired by John, while

Snickers had but one. Plus, Snickers had been "with" John first, placing her in the position of the established wife, with Samantha as the other woman. Whatever the reason, Snickers rarely let an opportunity pass for making Samantha miserable. She would snitch Sam's food, poke her in the ribs, tug her hair when she wasn't looking, and generally goad her into a rage. And when Sam obliged by throwing a tantrum and coming after her, Snick—never quick on her feet—would shuffle off, flapping her arms for John. Possibly because Snickers had mothered John's youngest, and possibly out of a sense of obligation because he had taken her to wife first, John would invariably side with her and turn against poor, put-upon Samantha in a rage and pummel her with his fists. She would shriek and hoot in anger, but there was nothing she could do.

Jessie was no less a trial for John. She kept pestering him. She would poke him, jostle him, swing on the bars just past his head. He took it for a while, but finally one day he'd had enough and he *exploded*. He gave an ear-splitting yell, bared his prodigious fangs, and pounced on Jessie. Jessie was naturally terrified. The two rolled about the floor in one big ball. He let out more frightening grunts; she, terrified squeaks. Alarmed, Snickers and Samantha put aside their differences to rescue Jessie from her father. As soon as they arrived, John let Jessie go and went to sit alone, brooding, in his customary place in the end cage. For his bravado, John didn't leave her with so much as a scratch. Jessie had drawn her father's blood with her fingernails. Nevertheless, when he passed by Jessie to change cages at the end of the day, she let out a scream of fright. "She was terrified that he would go after her again," said primate keeper Roseann Giambro. "But he didn't do anything. She has been good to him since. I think she needed that—a good whack."

The family feeling ran deep. Before Anaka, Snickers had had a baby named Justin, for Just in Time for Christmas, who was born in December 1983, but had died four months later of a bacterial stomach infection called shigellosis. "People say that gorillas don't have feelings," said Roseann, "but from the way that Snickers grieved for Justin I know they do. We had to take Justin away from Snick for two weeks before he died, to try to cure him. She was upset to have the baby taken from her, but it wasn't too bad. The morning that Justin died, she knew. She must have seen it in my eyes. She and John started moaning. It was a terrible sound. I tried to explain it to her, about the disease and everything, but of course she didn't understand, just looked at me and grieved. It was sad—terribly, terribly sad."

Snickers fell into a depression after that. For a long time she wouldn't take food from Roseann, and didn't snap out of it until Anaka was born the previous spring. "When Sam had her babies," said Roseann, "she'd watch and watch. I think she wanted to have one of her own." Finally, Anaka was born, and Snickers was overjoyed. She put Anaka close to the bars of the cage and let Roseann reach out to touch her feet as if to say, "See? It's a baby!"

All these gorillas had grown up together. Back in 1969 Dr. Snyder decided that the zoo needed to stop housing gorillas individually and start showing them in groups. In this, he was following the precepts of nature—since gorillas are highly socialized animals in the wild—and continuing the trend that Hagenbeck had begun in the nineteenth century. The Philadelphia Zoo had an enviable record of gorilla longevity ever since it instituted glassed-in exhibits for its primates. Massa held the record for longevity when he died in 1984. But through the sixties, the zoo had failed to persuade any of its gorillas to mate, or even to express a romantic interest in each other. Indeed, the zoo hadn't advanced beyond the ill-fated "wedding day" of Massa and Bamboo in 1935.

In fairness, most other zoos hadn't advanced either. But in 1959 the Basel Zoo stunned the zoo world by announcing that its female gorilla had not only given birth successfully (the second such birth; the Columbus Zoo had had the first), but had raised her young, something previously unimaginable. Dr. Snyder attributed Basel's success to its unique program of keeping gorillas together in a group from a young age. "The large apes are social and require a lot of training," he says. "Their young grow slowly, like human children, and that means they have a lot of learning to do. A young gorilla has to learn the structure of the group—grooming, tolerance, their place in the social hierarchy, the care of the young—all these things I believe are necessary for successful breeding." In isolation, however, that learning would be nearly impossible. As director of the Penrose Laboratory, he decided to undertake a research experiment. He obtained six baby gorillas from what was then still the Belgian Congo (soon to become Zaire) and other zoos. There were three males and three females altogether. Among them were John, Samantha, and Snickers.

Since the six gorillas were all about eighteen months old, Dr. Snyder realized that they would need some adult supervision. Ideally, of course, he would have liked to have them raised by adult gorillas, but he couldn't afford them, nor could he assure himself that they would "adopt" the children as their own. So he substituted some human females led by Ann Hess. "Our idea was to supervise the play and training of the six gorillas," said Snyder. "Everyone called the girls surrogate mothers, but that wasn't the idea. We were following some experiments in Michigan on the socialization of young monkeys and they were showing that the absence of the mother set these animals off on the wrong foot. The gorillas accepted each of the girls as another gorilla."

The young women were in the cage from morning till night, playing with the gorillas and teaching them everything they needed to know—even about sex. "Sure they played copulation," said Snyder. "Why not?" Essentially, the women found themselves running a rather wild preschool. They played with various toys, including a little red wagon, a clear plastic ball with a butterfly inside, and a calculating machine. One of them, Susan Pajkurich (who would later help with Dr. Snyder's woodchucks), went on *What's My Line,* where she stumped the panelists.

In the early seventies, however, Dr. Snyder's research project was canceled by Bill Donaldson's predecessor, Ronald Reuther, in an economy move. Without funds for the researchers, the project dwindled down to simply keeping the troop together and hoping for the best. A second male, a tremendous specimen named Westy, died, and the third was sold to another zoo. John was the only male left. Scrawny in his youth, he appeared at first to be impotent—much to Dr. Snyder's distress—because he expressed so little interest in procreation. Finally, he mated with a third female, named Haloko, to produce the baby Kiki; but despite the zoo's efforts, Haloko showed no interest in raising her, so Ann Hess took over the job.

John turned into a genuine stud and went on to sire the three other gorillas, Jessie, Chaka, and Anaka. However, it wasn't until this fall that he was allowed in the same cage with his offspring. Although that appeared to be a violation of the troop's original principles, Karl Krantz's predecessor, Dietrich Schaaf, had feared that John might harm his children if he was allowed to be with them. Roseann Giambro finally persuaded him to let John join them. "I was pretty sure he was not going to hurt the babies on purpose," she told me. "But I was afraid that he might hurt them by accident if he was annoyed with the females. He might have a fight with the females and the babies might get in the way."

The door to his cage was opened one day in the fall. The impetuous Jessie ran right over and planted a kiss on her father's face. Then Chaka toddled over and hung from John's arm. Snick ambled around next, with their baby, Anaka. "He just looked at the baby," said Roseann. "He didn't touch him until later that afternoon. I could tell that Snickers was nervous about it. Eventually, the two of them, John and Snickers, sat down next to little Anaka. He reached out towards her with his hand and he touched the baby very gently on her head. That was all." The two got used to each other gradually until now, in the spring, John would play quite freely with Anaka, but only in the privacy of the tunnels in back of the cage.

If the gorillas were melodrama, the chimps down the hall were burlesque. One of their favorite pastimes was to throw their dung at the keepers. "And they have pretty good aim," said Roseann. "I suppose it's funny when you see it happen to other people. But it's not so funny when it happens to you." Also, the chimps were ungrateful. If the gorillas don't like some of the special vegetables that Roseann gives them, they delicately put them aside. The chimps chuck them back in Roseann's face.

Bob Berghaier was trying his best to control them, but they were too much for him. The problem wasn't a lack of intelligence in the chimps, who are, after all, human's closest relations; the problem was the use to which that intelligence was put. During the spring, he'd tried to train them to go in and out of their cages at the right times by buying them off with little tidbits. If they left the cage when Bob wanted them to, he would hand them a piece of fruit. They soon learned to stretch out their hands for the fruit—while keeping one foot planted inside the cage to keep him from shutting the door. "They'll do anything to frustrate me," said Bob.

Smoke was the worst one. Physically he was your average chimpanzee—the long arms, lovable face, sleepy eyes. But he had been a laboratory animal (an ID number was tattooed on his chest); the zoo rescued him before he was put to sleep. Bob thought that life in the lab had really screwed him up. "He's like a weirdo you meet in the subway," he said. He was ostensibly Molly's mate, but he had a crush on another unattached female named Panny, short for Pandora. I was there one afternoon when Panny was in heat and Smoke was going wild with desire. A female chimpanzee's heat cycle is something to behold. Her private parts get so thoroughly engorged it looks as if she has either sat too long in a paint bucket, or is suffering from the worst case of hemorrhoids ever recorded. Pandora's nether parts protruded a good six inches, and they were driving the troop's two male chimpanzees into a sexual frenzy. Panny, however, only had eyes for Bob Berghaier, and she regarded Smoke as a terrible comedown. She wouldn't have anything to do with him, and this made Smoke seethe with frustration—and turn for satisfaction to Roseann, who was usually not far away. With his eyes focused lovingly on her, he would masturbate with his foot. He did this so often, with Roseann and other women, that there was a thick crust of semen on his leg.

As Bob and I watched, Smoke tried to get Panny's attention by swinging over to her and hopping up and down, but she pushed him away. Smoke then careered around the cage, swinging from bar to bar in a rage.

Meanwhile, in another part of the exhibit, Smoke's daughter Rookia, or Rookie, was sampling her feces as if they were cake batter. "I tell ya," said Bob, revolted at the sight, "you can dress 'em up, but you can't take 'em out." Not particularly liking the taste, Rookie scooped up a gob of the stuff, ambled over to the baby, Jane—named after Jane Goodall—and slapped the load on her head. Jane didn't know what had happened at first. She rolled her eyes up to try and see what had fallen on her, and then ventured a hand up to explore. She brought a finger down to her mouth . . .

The chimps wouldn't be going to the new primate center, but would stay here. No wonder.

As I was getting ready to leave, Bob revealed that he wouldn't be going to the new primate center himself. It wasn't that he longed to work in the new building—he'd made his feelings plain about that—but he wished to maintain his relationships with the animals, chiefly the gorillas, that would be going there. The selection of keepers was made solely on the basis of seniority, and Bob didn't make the cut. Obviously he wasn't pleased to be cut off from the gorillas and be left here with the chimps. He tried to put the best face on it, saying that the Rare Mammal House would be better in the future once the other primates were out, but his heart wasn't in it. "I'm younger than those other keepers," he said, referring to Bob Pittman and Giambro, who would be going. "I'll outlive 'em!"

——9——

It was a hot day in May, and Gene Pfeffer was hosing down the elephants. They changed color as the water poured over them, turning from dusty brown to steel gray. Jimmy McNellis watched him from the shade of the doorway. When Gene was done, he came indoors to light a cigarette and cool off.

Life had been calm around the elephants since Kutenga attacked him after Christmas. "Things are going good," he said. "Everything's been going good. I've been feeling as comfortable with the elephants after the incident as I did before. If anything, I feel better about 'em now."

"How's that?" I asked.

"I respect 'em more. I've always respected the elephants, but I didn't really know what they were like. It was, sure I respect the elephants, blah, blah, blah. All the time I'm thinking—they can't really be that bad. Now I know how bad they can be. When I got clocked by Kutenga, for the first time I knew what the elephants were about. I knew just how powerful an elephant can be."

Now that he had seen the worst, in other words, he could feel optimistic. "When will you know if you can go in with the elephants by yourself?" I asked.

"It's up to me. Management can say, go do it, but if I

383

don't want to go, I'm not gonna go." He took a puff on his cigarette. "But I suppose it's really up to the animals."

We strolled over to the large doorway to look out at them as they strolled lazily about the yard in the heat. The great hulks moved slowly through the sultry air as if they were passing through some denser medium. Occasionally, they nibbled at some piles of hay that Gene had thrown down. They didn't pay any attention to us.

"If they treat me badly," Gene went on, "then it's gonna take a little while longer. But I think I'm making progress. Jimmy and Danny are giving me more room. I'm getting better with the Asiatics—they take what I say. But the two punks are still giving me trouble. I've told Petal to get the hell away from the shovel, and she's backed off, so that's some progress. It used to be that when I hollered at her, she'd spit some water at me and act up. But you never know. You can feel like things are all right, but that's just their way of sucking you in. Just when you think you've got 'em conquered that's when they get you." He let the thought drift away. "But things are going all right.

"When I got hit, everybody said, 'Are you going back?' like they thought maybe I wouldn't," he went on. "Well, the thing is I never gave any thought to *not* going back. I never thought about getting hit. They said I must be crazy. Well, maybe I am, but I don't think so. It may seem scary to other people, getting whacked by an elephant, but it never entered my mind. That's just the way it is. A cop doesn't go out and worry about getting shot. He just goes out. He's got a job to do and he does it. I've got a job to do and I do it.

"I faced some scary things in the navy. But if there's a fire in the number two boiler, you don't worry about getting blown up, you go down there and put it out. I never worry. Those people who do the worrying, they're the ones up in the cushy offices. They don't do these jobs."

He looked out at the elephants once more. "You love them, don't you?" I said.

Gene recoiled at that. He hesitated to put a word on his feeling for the great beasts. *Love* wasn't it, certainly. "I respect 'em," he said finally. "That's what it is, and I hope that someday they'll respect me, especially Petal. She's the biggest. She does the most things. And she's the boss. Kutenga might think she's the boss. Kutenga wishes she was. But Petal's the boss."

"And someday you'll be her boss?"

"I guess. It takes a while. But the longer it takes, the more it'll be worth in the end."

"Any guesses when?"

"Beats me," he said. Then he gestured toward the elephants roving the yard in their stately fashion. "Ask them."

—10—

Throughout the spring, the workmen slaved to get the World of Primates ready for opening day, now set for June 7. Like characters in a speeded-up silent film, everyone dashed about planting trees and bushes, laying bricks, lining the moats with plastic, filling them with water, building the lemur exhibit, filling it with lemurs, installing the marmoset exhibit in the adjoining orientation center, adding marmosets, laying down hot wires, building the trellis over one viewing area, painting, raking, soldering, polyurethaning, finishing off the primate bedrooms, installing the gibbons, orangutan, drills. . . . Miraculously, in their determination to complete the project on time, the workmen overlooked their union designations and helped out with whatever was needed. Maintenance men planted shrubs; laborers hauled mulch. Rick Biddle thought it was marvelous.

Toward the end of May, the center was finally ready to receive the gorillas. Everyone had been dreading it for months. Difficult to deal with because they are so powerful and so clever, gorillas are also priceless. They are impossible to obtain from the wild nowadays, and zoos won't sell them. Yet every phase of the move was potentially life-threatening: darting them, hauling them out on stretchers, giving them checkups on the Penrose Annex operating table, then carrying them over to their new quarters in the primate center and hoping they didn't go nuts when they woke up.

Everything had gone smoothly with the other primates. The gibbons had been moved from their old haunts under the monorail into their new quarters, filled with ladders and ropes for climbing. They immediately started flying about the new space as though they had lived there forever. And the orangutan Bong was the same as ever. He had been placed in the tunnel that led to his indoor cage several days ago in hopes that he would move into his new bedroom when he was ready. He had woken up from the anesthesia, but he hadn't budged. He still filled the barred tunnel like a boulder overgrown with reddish, long-haired moss. The drills were doing well. Wilbur's hypersexuality had even calmed down a bit in his new environment, although it was too soon to tell if the change was permanent. The lemurs—slender, bushy-tailed primates from Madagascar—were obviously happy in their wire-mesh exhibit at the end of the building. One of the females had staged a housewarming by giving birth to twins.

Now came the real test: the gorillas.

It was nine o'clock on the morning of the big move inside the Rare Mammal House, and Samantha was screaming. Frightened yips and eerie, doglike barking ascended to ear-piercing shrieks; it was an aria of outrage and fear. The sound reverberated through the building and brought to mind all the terrors of the jungle. The keepers had shifted her into the tunnel behind her cage a few moments earlier. Now she spotted Keith and she knew for sure what was coming next.

Inside the keeper's room that leads to the tunnels at the back of the animal cages, a whole crew of worried-looking staffers was milling about. Roseann looked the worst—drawn and tired. "I haven't been sleeping too well," she said above the din. She seemed to wince at every blast of sound.

I nearly had to cover my ears when I got to the damp corridor that reaches the tunnels. It was like a passageway into the hold of an oil tanker—water dripped in places, and rust stained the painted walls. In the tunnel, walled with iron bars, Samantha was crouched on all fours screaming her lungs out, fangs bared. Chaka took cover behind her. From the doorway,

a zoo cameraman—there to record the move for the zoo's archives—shone a bright light on the animals as his video camera
rolled. Keith and Mike stood together, looking grim.

As the senior veterinarian, Keith was in charge of the move,
and he seemed particularly wary. The last time he'd had to
dart Samantha, Jessie had jumped him. He had left the door
to the adjoining cage closed but unlocked for a few moments
while he attended to Sam, and Jessie had burst through it. She
must have thought that Keith had just killed her mother. "She
bit me on my shoulder and scratched my face and ears," he
told me. "Then she ran out into the hallway. There were three
or four other people out there, and Jessie was cruising around.
Finally, somebody managed to throw her into her cage. She
was smaller then—maybe sixty pounds. She must weigh eighty
pounds now. They'd never be able to hold her." Keith wouldn't
make the same mistake twice.

"Ready?" Keith asked Mike.

"Yeah," said Mike. "Are you?"

"Let's go."

Keith prepared a dart for his blowgun (quieter than the
pistol, and less stressful to the other animals in their adjoining
cages), then climbed up on a ladder to peer in at Samantha
and son. Samantha quieted for a moment. She seemed to be
sizing up the situation just as Keith was. Then she came to the
only possible conclusion and started screaming again louder
than ever.

Keith drew a bead on his prey and fired. There was that
familiar *fft* sound as the tasseled dart sped toward its target.
Then a *whump* as it hit the flesh by her right shoulder. Almost
instantly Samantha reached up to grab the spot with her hand.
She wailed some more and, in a frenzy of pain and shock,
shook the tunnel as if she might pull it loose from its supports.
Keith returned to reload.

Her screams by now were earsplitting. "This is nothing,"
said Keith. "You should hear her when she's really worked up."
Spotting Roseann, he said, "Chaka is really holding on. It's
going to be hard to get him out of there."

"I'll go get a banana," she said, and hurried out of the
room.

Samantha plucked out the dart and tossed it to the cement

floor. It fell with a clink. "It's easier with cats," Mike said. "They don't have hands." Still, the dart had done its work. The grunts and yowls softened. Keith climbed back on the ladder and fired another dart; this one smacked into her chest. Gradually, Sam slumped onto her back. Chaka was terrified. He jumped up on her and scrambled frantically around her body as if searching for some sign of life.

Roseann came up to him with a banana. "Chaka," she said soothingly. "Chaka sweetie." She extended the banana toward him. Chaka screeched and hurried to the far side of the tunnel. "Don't you want a banana?" she asked. Chaka looked at it with a child's eyes, mildly tempted, then pulled back and looked around, blinking. Roseann pushed the banana closer; he pulled his nose away. No. He started to yell in a boyish soprano.

Roseann pushed a hand up through the bars and onto Chaka's back. He froze. She pressed another hand through the bars to caress him gently like a mother with a newborn. Not yet two, he barely weighed thirty pounds and was still too small to be dangerous. Chaka fell silent; he was just breathing now. Roseann tried to pull him to her, but he clung to his mother's body.

"He's so scared," Roseann told Keith. "He's never seen so many people."

"Move back, everybody," said Keith.

The keepers who had crowded into the passageway to watch pulled back inside the door now.

Chaka moved off his mother and retreated deeper into the tunnel. Roseann reached out her hands still farther, imploring him. He climbed back onto Samantha and held on to her prostrate form.

Roseann's efforts were of no use. Chaka would have to be taken by force. At a signal from Keith, a keeper appeared with a net, and staying as far as possible from the slumped Samantha, he brought the net down on Chaka and then dragged him to the bars. There, Keith injected him with a tranquilizer, and almost immediately he fell still. Keith didn't dare use the blowgun on such a small animal; he might miss and hit an eye, a joint or a vital organ.

Now that both animals were subdued, a door was opened to the tunnel. Chaka was pulled out and placed in Roseann's arms. Roseann cuddled him like a sleeping toddler, his chin resting on her shoulder. Then Samantha was dragged out and placed on a stretcher of loosely woven rope. About a dozen keepers grabbed various handholds and carried her over to the Penrose operating room. The Rare Mammal House fell silent. Jessie would be next; John, Snickers, and Anaka would go tomorrow.

At Penrose, Samantha lay sprawled out on her side on the operating table. Her body was sinewy, but her face was gentle. Her long arms stretched over the edge and her hands hung down. They were extraordinary: the fingers thick as bananas, tipped in tough nails, the skin hard and crusty as old leather. Gradually, the room filled with the locker-room smell of nervous gorillas. Roseann stood at the foot of the table with Chaka in her arms. "He has a death grip on me," she said. "He used to be a little baby, now he's a gorilla. Smells like one, too"

Mike Barrie went to work on Samantha. He inserted a thermometer up her rectum, gave her some shots for tetanus, diphtheria, and rabies in her buttocks, and then tied off one arm to take a blood sample. Mike shaved the underside of Sam's forearm, then wiped it with alcohol before inserting a needle.

Carl Tinkelman peered in Sam's mouth. Her teeth were as thick and yellow as tusks. "She's got some nice fangs," he said. The incisors were of vampiric dimensions, at least an inch and a half—her bite would be much worse than her bark. Then he looked closer. "Uh-oh, one of her front teeth is missing."

"You know they play rough," said Roseann.

"Any cavities?" asked Keith playfully.

"Nope," said Tinkelman. "Looks like Sam has been brushing every day." He scraped off some of the plaque.

Roseann put Chaka down on the floor for a moment, but despite the tranquilizer he started to scream, so she picked him up again. "I didn't know he could tell," she said sheepishly. She sat down cross-legged on the floor and held him in her arms.

"How's it going?" I asked Mike.

"You can tell by the level of excitement," he said. "If it weren't going well, we'd be moving quicker."

Karl Krantz came in with some ink for taking Sam's footprints. He needed the images for an educational display. After smearing one foot with ink, he pressed a sheet of paper to it. He took several impressions, then wiped the foot clean with alcohol. Finally, he blew on the foot to dry it.

"Yeah," said Mike with a smile. "There's our lips-on curator."

Karl then measured Sam's armspan (six feet two), hand length (nine inches), and chest circumference (three feet, seven and a half).

"You should send them in to Miss America," said Mike.

When they were finished, they lifted Samantha onto the stretcher again and carried her, feet first, to a waiting pickup truck. It drove slowly, like a hearse, to the new primate center. Some workmen had to move a wooden barrier to let the truck go down to the entrance.

The center wasn't quite unfinished, but I could see it came about as close to a nonbuilding as an architect could get and still keep his self-respect. The breezeways pulled you in clear through the building to some viewing balconies on the far side. There, you could see the lush islands were the primates would soon romp. The rest of the building would only service the outdoor exhibit—holding the animals after hours, providing facilities for the keepers. It was a backstage; the outdoor exhibit was the stage.

The gang of keepers hustled Sam into one of the tunnels in a back room. These were the tunnels that Jim Bradberry had fussed with so much in his architectural renderings. They were great shafts of aluminum that ran across one side of the room at head height, giving the animals passage between the outdoor exhibits and their private bedrooms. The keepers scoffed at them, but they did their job. The door to the tunnel was opened, and Samantha was more or less shoved in. Then Mike Barrie climbed into the tunnel himself to reposition her arm so she didn't cut off her circulation. Roseann had carried Chaka down in her arms, and she placed him inside the tunnel, too, next to his mother. The tranquilizer had worn off by now. Seeing his mother prostrate beside him in the strange tunnel, Chaka shrieked and shrieked. But there was nothing anyone could do.

Wilbur Amand came down to check on the animals a while

later. Chaka had quieted down, but he still looked mournful. Wilbur tried to comfort him through the bars. "Not only do they knock you out, but they put you in a whole new building," he said. "But it's okay, it's okay." Chaka watched him in silence, disbelieving, still sitting beside his sleeping mother.

Jessie was next. She was darted more calmly. Keith carried her in his arms into the Penrose Annex.

"Getting your revenge, Keith?" asked Eileen. She was remembering his run-in with Jessie the previous year.

"You know I wouldn't do that," said Keith lightly.

"Take it from me," said Mike. "Jessie's in very good health."

"Except for her fingers," said Keith. She was missing two of them.

"Somebody bit them off," said Mike. "Right down to the bone. We suspect Snickers."

As they had with Sam, the vets and Tinkelman did a blood test on Jessie, a teeth check, and a rectal culture as well as giving her various shots. Then Keith carried her down to the exhibit and placed her in a tunnel. She and her mother looked like teddy bears put in for Chaka's comfort. But Chaka would not be soothed.

A little later Eileen was back in her lab. At first I though that she had a different hairstyle, but then I realized that there was a tiny quivering marmoset in her hair just over one ear. "His name is Moo," she said. "I guess he's kind of a consolation prize," she said.

Ann Hess was back. It turned out that she didn't have to have surgery on her finger after all. At the last minute, a doctor had told that she might be able to restore some of the motion in her index finger with some physical therapy. So she had returned to the job, and she had reclaimed the A-team.

"Moo's not a binturong, but he's okay," said Eileen.

The world's tiniest primate species, marmosets look like shrunken versions of the scary monkeys in *The Wizard of Oz*. Eileen's tiny Moo could sit comfortably in a tablespoon. Marmoset mothers can only handle two infants at a time; Moo was the third.

"These are rare little guys," she said, warming to the subject. "The Philadelphia Zoo is one of only two zoos to have them. Marmosets are pretty smart. Moo knows me already. He jumps into my hand." She set him down on the lab table to demonstrate. It was a little like watching trained fleas. But, sure enough, Moo hopped up onto her finger. "See, he doesn't do that for other people."

He did some grooming now that he was in the public eye.

"When his mouth gets dirty, he wipes it off. He also runs after Roseann." She tickled Moo's chin. "You don't have to use darts to move these guys."

"Lucky thing," I said.

"I'll say. Did you hear that screaming with Samantha? I thought the building was going to fall down. Even Keith was scared. He might have looked calm, but I could tell. But if you want fear, wait till tomorrow. Wait till he does John."

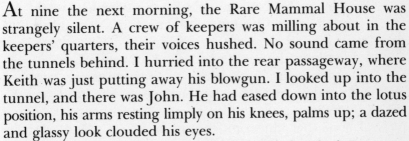

At nine the next morning, the Rare Mammal House was strangely silent. A crew of keepers was milling about in the keepers' quarters, their voices hushed. No sound came from the tunnels behind. I hurried into the rear passageway, where Keith was just putting away his blowgun. I looked up into the tunnel, and there was John. He had eased down into the lotus position, his arms resting limply on his knees, palms up; a dazed and glassy look clouded his eyes.

"You been teaching him yoga?" Eileen asked Roseann. Roseann smiled wanly and kept her eyes on John.

Keith had decided to dart him ahead of schedule, before too many people—administrative staff, keepers—arrived. Usually, a smile is never very far from Keith's lips, but now he seemed deadly serious and a little shaky as he turned back to confer quietly with Mike Barrie.

Eileen was right: John had scared him. I pushed my way over to her to find out what had happened. "It went fine," Eileen said. "But it's just that John is so intimidating. One slam of his hand against the bars and the whole zoo shakes. We're talking *big and strong*. Keith was white. I'd never seen him like that."

But King Kong had fallen. Several keepers were pulling

on him to slide him out of the tunnel. Chuck Ripka was bent over to take the brunt of the weight on his back when John toppled out.

"Everybody ready?" asked a voice. "Set? *Heave!*" And out came John onto the thick rope netting. Four or five keepers were ringed around the net, Chuck Ripka underneath. Chuck's face nearly turned purple when the four-hundred-pound load down on him. John sank lower and lower, and Chuck nearly disappeared. But finally the descent was halted and John's massive body, with its thick muscles and long arms, was eased out through the door and over to the Penrose Lab.

Keith had recovered his good humor by the time John got to the table, and his gaiety infected his colleagues. He started poking through the matted hair around John's underparts to do a rectal exam. "It should be in here somewhere," he said.

"You mean he doesn't have an anus?" asked Mike.

"No wonder he's so big," said Keith under his breath.

"Help me spread his buttocks here, Karl," he said to Karl Krantz.

"I don't know him well enough," Karl replied.

Samantha had certainly been sizable once she was stretched out on the table, but John was enormous. His head was the size of a medicine ball, and it was crowned with a thick sagittal crest of bone that ran from front to back like a Mohawk haircut. His chest was twice as big as the average man's, and his muscles would make a weight lifter swoon in envy. "God, he's gorgeous," said a lab technician who was looking on. Roseann slipped around to stand by John's head. She stroked his hair. "My honey," she said. Then she tickled his ears.

Aside from these occasional jokes by people who had worked with John before, the room was much quieter than yesterday with Samantha. People were awed by John. Even unconscious, his eyes staring vacantly at the ceiling, he exerted an immense power, and he commanded the room as he slept. Some of the keepers produced cameras to take photographs of themselves standing next to the big gorilla like tourists at a national monument.

Once again, Karl Krantz was busy with his tape measure,

recording John's canines (one and a half inches) and feet (thirteen and a half inches).

"I make that about a size sixteen E," he said. "That could be useful," he added, looking down at his own small feet.

"Yeah, but you'd be embarrassed to take your shoes off," said Mike.

Now Karl measured John's armspan (eight feet one), biceps (nineteen inches), neck (thirty-three inches), and chest (four feet ten).

"That's better than Carol Doda," said Keith, referring to the busty actress.

"Yeah, but you didn't have him on your back," said Chuck Ripka. "He's a load."

"But he solves the Eagles' middle-linebacking problems," said Bob Berghaier.

"I wonder what John is thinking, Bob, as he hears your voice," said Roseann.

Finally, the work was done, and the keepers resumed their positions to cart John out to a waiting truck and then down to the primate center.

Already the new center had suffered its first mishap. In order to fit the gorillas' tunnel in, Jim Bradberry had had to drop it straight down three feet at one juncture, not far from where Samantha had slept off her anesthetic. She had woken up late yesterday evening, and stumbling about in a daze with young Chaka on her back, she had wandered too far down the tunnel and dropped like a stone off the ledge. Fortunately, the two gorillas had come out of it undamaged, and the staffers had treated the incident as a joke—all except Roseann. She had seen it happen, and she had been terrified. To her, this was more evidence of the dangers of the new building for her gorillas.

Samantha and Chaka had moved into their bedroom by now, so the tunnel was empty. The vets decided to put John in at the lower level at the far end rather than risk his taking a tumble as Samantha had.

"How's he going to get up to the top level?" asked Mike, noting the steep climb John would have to reach his bedroom.

"Like this," said Keith, and he suddenly jumped into the tunnel and then nimbly climbed up to the higher level using his hands and feet. As gracefully as he moved, he was too tall for the tunnel, and he had to squat uncomfortably to keep from banging his head. It was awkward to see him so tightly confined, his universe reduced to an aluminum box. It made you question the whole premise of putting any living being in a cage.

"Yeah," said Mike, "but you're using your hands."

"He has hands," said Keith.

"Oh, right," said Mike, playing dumb.

The keepers put John in. He looked like an eagle squeezed into a canary's cage. Still dazed from the anesthetic, his eyes looked as dull and unseeing as a stuffed animal's.

Samantha and Chaka had been moved into their new bedroom. You could see it from a terrace by the old Kangaroo House, now the primate center's orientation building. The room might have come off a large submarine. It seemed nautically tight, and had lots of poles and ladders leading from floor to ceiling. A few thick ropes hung down for the gorillas to swing from, and there were a couple of those springy PVC-coated hammocks Jim Bradberry had labored so hard to create. Chaka had taken to the place as only the young could—toddling around and investigating everything. He showed off a couple of nifty hand-over-hand maneuvers on the ropes. Samantha, still groggy perhaps from her fall this morning, was gingerly knuckle-walking about the floor, checking the boundaries of her new domain.

Wilbur stood at the glass. "They had a tough day yesterday," he said. "I felt for them all." We watched the two gorillas in silence.

Then it was Snickers's turn. She went under without a struggle. Tiny Anaka came out in a net. Tranquilized, she was curled up in Roseann's arms.

After the keepers laid Snick's tubby body out on the table, Dr. Tinkelman discovered what may have been making her so cranky for the past few months: her teeth. They looked horrible—nubby and brown-stained as though she had been a lifetime smoker. One bottom tooth had rotted away so badly the nerve was exposed to the air. "Poor kid," said Tinkelman. "She must be in constant pain." Tinkelman immediately set to work to extract it.

It was an excruciating procedure. Human teeth don't come out all that easily, but gorilla teeth are at least twice as big, and they are fused to immense jawbones. It would be like trying to pull out a headless spike from a four-by-eight. Tinkelman began with common pliers, latching onto the tooth and then pulling with all his might. The pliers made a ghastly grinding noise as Tinkelman strained. After a few minutes of this torture, Snickers swung her arm around toward him. "Keith!" Tinkelman yelled. "More ketamine!"

Tinkelman went back to work and made some more crunching noises. Blood dribbled out of Snickers's mouth and down her chin.

"I've got to loosen it from underneath," he said. With his scalpel he sliced deeply into the gums and folded rubbery gum tissue back in her mouth to get it out of the way.

"To get this one out, I'm going to need a hammer and chisel," he said.

A photographer laughed.

"It's no joke," he said. "That's what Dr. Amand had to do with Massa."

For the moment, he attacked the tooth with a "gouger" that looked like a cobbler's awl, jamming the instrument into the gums around the tooth trying to break it loose. At last some enamel popped free.

"Finally," I said.

"Yeah, but we're not done," said Tinkelman. "That's just a little piece."

Snickers growled through a flood of saliva in her throat.

"I know, I know," said Tinkelman in response. "Poor Snick."

"Like pulling teeth, isn't it, Doctor?" asked the photographer.

Tinkelman ignored him.

Bill Maloney strolled in. He'd come up from the primate center.

"How's John doing?" asked Mike.

"He's all right," Bill answered quietly. "He's coming around."

"What did he say?" someone asked.

"He said, 'Run for your lives!' " Mike yelled out. " 'John's bending the bars!' "

Another doctor came in, a rheumatologist from the University of Pennsylvania named Ralph Schumacher. Karl Krantz had asked him to take a look at Snickers's arthritic elbows and knees. Karl himself had consulted Dr. Schumacher for some stiffness in his hand. While Tinkelman continued to hack away loudly at Snickers's teeth, Schumacher, a middle-aged fellow in a white coat, moved in to examine her arthritic joints.

It was interesting to watch the rheumatologist work. Much as Keith and Mike cared about their animal patients, they treated them somewhat roughly when the animals were under anesthetic. Dr. Schumacher, used to dealing with humans, handled Snickers's elbow as though it were made of fine china. Holding the forearm lightly in his fingertips, he ever so delicately eased the joint back and forth, feeling for the stiffness and grinding of the arthritis the way a safecracker might listen for the fall of the tumblers in a lock. He had such presence, the whole room fell silent as he worked.

Just then, Tinkelman's pliers snapped off in his hands. He muttered an oath and reached for a second pair, and resumed his labors. "They're just not made for a gorilla," he grumbled.

Dr. Schumacher could draw no immediate conclusion. But he didn't believe there was a connection between Snickers's tooth decay and her arthritis, as one of the keepers suggested. Tetracycline, the medicine Snickers was taking for her arthritis, wasn't known to cause cavities.

Seeing that he had made so little progress, Tinkelman reached for the heavy artillery: a power drill to annihilate the tooth.

As he brought the drill down to the tooth's stump, Snickers stirred again and groaned softly. Everyone jumped to secure her arms and legs while Keith gave her another injection.

"You know, Keith, you should pose for a picture with Snickers and Anaka," said Bob Berghaier, "Philadelphia couple makes genetic history."

Finally, Tinkelman managed to free the last hunk of tooth.

He held it up with his tweezers triumphantly. Then he stitched Samantha's wounds with a giant needle that looked like a bent fishhook. "There," he said when he was done. "All better."

With that, the keepers carted Snickers off to her new home.

—12—

The fiscal year was also winding down, and Rick Biddle was putting together the final numbers up in his office, lined with pictures of sunsets. He'd just come back from a brief golfing vacation with his wife, Ann, on Kiawah Island off the Carolina coast. "*Great* sunsets over the beach," he said. He seemed to be drooping a little.

He reviewed the year for me. "From an attendance standpoint, the period from October to March wasn't too bad," he said. "Except we got killed in February. The weather turned on us, and we got some really cold weekends. Valentine's Day, our adopt-an-animal day, was freezing. We usually pull eight thousand and we got three thousand. That wasn't too good. We'd budgeted forty thousand for the month, and did twenty-one thousand. But we pulled out of it in March.

"See, the key to our year is when Easter falls. Our peak season begins on Easter weekend, and we were lucky this year that Easter fell on March thirtieth. The parochial schools are off the week before, and the public schools off the week after. We'd budgeted fifty-five thousand for the month and we got down to the weekend of the twenty-second, and we were sitting with forty-two thousand, and I though, 'God, are we in? Are we gonna make our budget?' Quite frankly, I was worried about it. I told Scott, 'I think we're screwed.'

"Then we got a break. On Easter week the weather turned beautiful. The temperature shot up to something like seventy, seventy-five. On Thursday, Good Friday, Saturday, and Easter Sunday, we drew forty-five thousand. I couldn't believe it. We wound up doing ninety-six thousand five hundred for the month."

He showed me the figures on previous Marches, all of which hovered around forty thousand. "See," he said, "there was a real swing there."

Since then, the count had held to projections, with April dropping a touch below, and May rising a little above.

"When this year started, I said that if the koala brought in an extra hundred thousand people, we would finish up with a million three," Rick went on. "Some people around here thought I was drunk when I made that prediction. But if June holds, and I think it will, we'll end up with it almost exactly. I'll be damned if I wasn't right on the money."

At that turnout, the zoo would be up sixty thousand over last year, which itself had been a record setter. With each person spending, on average, $5.50, that meant a surplus of $330,000 for the year, which would be put back into the deferred maintenance fund. And with the primate center, it looked as though the zoo would get off to a good start through the summer.

"I think the World of Primates is going to go hell's bells," said Rick. "I mean, it's such great publicity. To open the morning paper and see John standing there"—he was referring to an *Inquirer* feature on the new exhibit that had run that morning—"I think it could be Katie bar the door."

But a cloud was darkening the zoo's sunny financial picture. "We've got a little problem with our liability insurance," he said. He'd just gotten the bill. After climbing from $160,000 to $253,000 last year, the annual premium had jumped to $720,000 this year. "I knew it was coming," he said. "But it's still blown us out of the water. See, there's no room to maneuver—we need the insurance to open." Even though the zoo had submitted only one major claim—for a child who had slipped on the floor of the TreeHouse and sprained an ankle—it was still caught up in the general escalation of liability rates. So Rick had to cut his capital expenditures back to zero, although he had received a million dollars in requests for everything from new animal exhibits to new electric carts. He'd had to trim individual department budgets back to the bone. It also made the prospect of the Schuylkill Expressway's closing for repairs over the summer even more frightening. Right now, Rick had pushed the projections up so high, and pulled the budget down so low, that there wasn't much room for any shortfall.

He felt sure that a shortfall would come. He'd been so lucky for the last few years. Lucky that the weather had been good on important weekends, cool enough on Labor Day to keep Philadelphians from going to the beach instead of the zoo, warm enough on Easter to bring the visitors. Easter would fall nearly three weeks later next year, on April 19. It had only been two years since the zoo had broken 1.2 million, and now he was counting on breaking 1.3 million. Everything had to fall exactly right for that to happen, and Rick didn't believe that his luck could hold. On the bright side, the zoo had pulled in the work on the Boston zoos and the Camden aquarium to boost its budget. But even so, he had to show a $300,000 deficit for the coming fiscal year, and Rick didn't like that at all. "You run and run and run and only stay in place," he said. "If that."

"Is Bill worried?" I asked.

That brought a smile. "Bill never worries. I told him about the insurance thing at the end of a meeting with him and Scott Schultz last week. I said, 'Bill, the liability insurance came in a half million over last year.' And you know what he did? He got up and starting walking out the door. And he just kept walking—right down the stairs. So I kept talking, and I said, 'And that's not including the one hundred and fifty thousand dollars we're increasing my salary, the fifteen thousand-dollar car we're giving to Scott Schultz, and the eighty thousand-dollar travel budget for senior staff . . . ,' and on and on like this. Finally, Scott and I just rolled on the floor laughing. But that's Bill's way, he doesn't worry. He never worries. He lets us worry."

—13—

The official opening for the public was set for June 7, but the major donors' preview on June 3 promised to be a bigger deal. The mayor would speak, there would be cocktails, the Main Liners would come, and all the zoo people would be there.

Most of the zoo dignitaries arrived early, and were clustered about the viewing areas around the new exhibit. Binky Wurtz looked especially proud. Volunteers had been passing out V.I.P. buttons, for Very Important Primate, and Binky was wearing two of them, one on each lapel. And Bill Donaldson's smile was even wider than usual as he led around some three-piece-suited major donors like Bob Smith, president of the Pew Charitable Trust. With all the well-dressed people hanging about, I didn't notice right away what they were all staring at: The primates were out in the exhibit. They were a beautiful sight.

As I looked out from the main viewing terrace between the animal bedrooms and the orientation center, I could see John lumbering around one of the low hillocks, reaching up into the branches to strip off some leaves and shove them into his mouth like potato chips. "It's been a long time since I saw a man like that!" one woman exclaimed. Samantha sat on the ground investigating her toes, while Chaka toddled about in the grass. And Snickers, all recovered from her dental work, shuffled along with Anaka clinging to her belly. Jessie swung from a branch of an oak.

Everything looked so natural—the way the sunlight fell on their shoulders and the breeze tossed their fur. These animals were no longer purely exhibit specimens; they were part of the natural world. And so were the viewers. For the first time (except when the animals were anesthetized), no bars or glass intervened between me and the gorillas. I could see them clear and sharp in the natural light, and I could hear them as they knuckle-walked across the grass, thrashed about in the trees, and grunted contentedly.

On the middle island, the gibbons had turned the high sycamores into a jungle gym. In a twinkling, they leaped from the ground onto the lowest limb, then climbed hand over hand to the topmost branches, fifty feet in the air, where they swung back and forth like the most brazen gymnasts. One gibbon found a squirrel's nest up there, and he was having a lovely time ripping it apart like a New Year's piñata. The sight must have gladdened the heart of Chuck Rogers. When the gibbons did touch the ground, they tiptoed through the grass with their hands daintily in the air like maidens crossing a dewy field. On the farthest island, the drills stalked about—their dark faces looked like sculpted ebony, and their rear ends bore an iridescent sheen.

Only poor Bong, the orang, was unhappy. A mass of reddish curls, his great fleshy face curved into the saddest imaginable frown, Bong hung in the low rectangular doorway to the middle island he would share with the gibbons. He looked like the kid who can't swim watching all his friends enjoy themselves at the pool. He simply couldn't bring himself to come on out and try it.

Jim Bradberry was showing his wife around the exhibit. He looked especially dashing in an Italian suit. He had been here yesterday when the gorillas first stepped out into the exhibit. "Samantha came out first," he said. "It was like *2001,* or the beginning of time. It was awe-inspiring. All of a sudden, there she was. The wind was coming through the trees, blowing her fur. The birds were chirping. She was *outside.* I think she liked it. She kinda sniffed the air and went over to the trunk of one of the smaller trees and looked up. That was something. I don't think she'd ever seen a tree before. She seemed to like the look of it. She patted the bark with her hand. Then she went up that little hill over there, and the sunlight shone on her. We were all watching. No one said a word. Everyone was holding their breath. It was so beautiful.

"But the gibbons were the best," he went on. "They took to the place like they'd been here all their lives. The keepers put them in a few days ago, and they wouldn't come back inside at all for the first two days. They were having such a great time running up and down in the trees. But that's okay. They can stay out if they want to."

Like a kid showing off a new train set, Bradberry took me around the exhibit to marvel at it from all angles. The things that had worried him still worried him, he admitted. He didn't like the looks of the hot wires lining the moats. And he was sure that many of the plantings wouldn't make it through the summer. The grass, which had only been in since April, was already getting trampled and uprooted. And John had stripped some of the small trees. "He treats the place like one big salad bar."

Also, disaster had struck. One of the large panels of glass in the porchlike viewing area at the far end of the exhibit had been smashed: It looked like a shattered windshield. A metal sheet had been attached to the interior to protect the animals from loose shards of glass. Arlene Kut had gone to work to minimize the PR fallout by putting up an explanatory sign. OOPS! it said. This was the very thing that Jim had worried about most, and now it had happened. "I don't know how they are going to fix it," he said. "One of the workmen knocked into it. It was a total fluke that it broke. I felt so sorry for the guy."

This viewing area was set down into the earth, and it was noticeably cooler as a result. The gorillas were considerably higher up. That was intentional. "The animals are respected more if you have to look up at them," said Jim. "That's one of the things we found out visiting other zoos. And it makes sense, if you think of the old bear pits. People would look down on the bears and spit on them. That was terrible. This is much better." The moat pooled the water into a lagoon around one corner, out of sight. "The lagoon extends the space," he said. "An old Disneyland trick."

As usual, you couldn't see beyond the exhibit. All sight-lines were concentrated on the moated islands and hillocks. "We control what we can," he said. "We want you to look in here. With earth berms and plantings we want to wall you off from the world. We want you to feel that when you come in here, you have come into a whole new environment—the World of Primates."

We sat down on a bench under the trellis, and Jim let out a long breath. "It's been four and a half years for me. So many things have happened. Mary-Scott has had two children. One of our stone masons on the job, John Conti, committed suicide. He'd taken over the family business, and finally the pressure got to him. There's a plaque here to commemorate him." A small metal plaque had been set into the stone wall over our heads. "And Dave Wood's predecessor, Rick Beyer, died in a car crash. A project manager had a heart attack. I got married. Lots of things happened."

Jim looked out at the exhibit once more. "It's nice," he said. "And it's going to get nicer."

"Hey, you hear about our great bear hunt?" It was Keith Hinshaw. He came up, slapped me on the back, and told me all about it. A black bear had gotten loose in upper Chester County, about twenty miles from the zoo. "I guess it was terrorizing some people," Keith said as though he couldn't imagine why. Some game commission wardens had shown up to dart the animal, but the anesthetic hadn't taken effect. The wardens called the zoo for advice. Keith and Mike agreed to help, so they were whisked to the scene by helicopter, and soon spotted the animal charging along in an open field with a vast horde of humanity chasing after it. The helicopter touched down, and Keith sprang out, anesthetic gun drawn. He ran over to get within range. "But as soon as I got there, the bear keeled over, dead asleep," he said. Keith did a physical on the animal, found him to be in good health, then helped the wardens move the creature deeper into the woods, and released it. There was nothing to it. "He was just a bear who went for a walk and didn't stop in time," he concluded.

Keith lightened up now that the primate center was in place and the gorillas were safely moved. Amazingly, the entire move had gone without a hitch, with the possible exception of the lemur female's unexpected delivery of two babies named Androcles and Andromeda. "I knew there had been some mating behavior," said Keith. "And I was pretty sure she was pregnant because I had palpated her and I thought I'd felt a couple of heads. But I didn't know she was so close to delivering. But then after two days here, the babies just popped out. Stress can induce labor in lemurs, just like in humans. But the babies are doing fine."

As for Bong, Keith wasn't concerned. "He's just taking his time," he said. He explained that Bong had never been out since he was caught in the jungles of Borneo as a baby. He was twenty-five now. "But he'll get with it."

Mayor Wilson Goode suddenly showed up to cut the ribbon that officially opened the building. He pushed through the crowd like a heavyweight being led to the ring by his entourage. Goode once asked Bill Donaldson to be the city's managing director in charge of the fire and police departments, but Donaldson turned him down because he didn't want to leave the zoo. The two seemed cordial now, but not the best of friends.

The crowd hushed and gathered about the central viewing area to hear the speeches. Donaldson welcomed everyone in a high, soft voice that didn't carry well. Then Goode boomed out a speech that everyone could hear about the importance of partnerships between public groups and private ones for just such ventures as this primate center. "I'm pleased tonight to cut a six-million-dollar ribbon," he exclaimed as he snipped. Everyone clapped. Afterward, he posed for the photographers with his arm around the sculpture of Massa, which was now in its final resting place on the terrace.

It was a beautiful evening. The sun was just dropping over the elephant house and sending long shadows across the primate center. A bossa nova band struck up "The Girl from Ipanema," and everybody surged toward the cocktail table. I preferred to take a stroll. I swung around by the rhino yard for a last look at B.J. He was trotting friskily around his mother, Xavira, idly munching on some hay when I pulled up. He stopped to gaze at me for a second, his snout grazing his mother's hind leg, his donkey ears twitching. He was growing up fast; before long, he'd be shipped out to another zoo. I waved at the two of them. "See you," I said. With a toss of the head, B.J. strolled off to a far corner of the yard; Xavira continued to eat.

The sounds of the music fell away behind me as I passed the TreeHouse. By the time I reached the muntjac yard, I couldn't hear anything at all except the wind through the trees. The little muntjacs, the small deer with pointed ears, were still out. They scurried about their pen and tumbled over each other.

In Wolf Woods only a few tufts of winter fur still clung to the wolves' backs. The three sisters were settled in their usual configuration about Taboo. All but Trillion pricked up their ears and looked at me; Trillion, contented now, continued to doze. The three others stared at me as though they had seen me somewhere before. In the evening light, their eyes glowed a strange neon green. I spoke to them quietly, afraid that other wandering zoo guests might overhear. "I'll miss you," I said. I walked off toward the cheetahs and barasingha deer, but before I was out of sight I turned around for one last look. The three wolves were still watching me, still intent.

The alcohol, or something, had started to take effect on the crowd when I returned. Rick Biddle greeted Scott Schultz with a triumphant high five. "Well," Rick exclaimed, "we pulled it off!"

"The crap was really hitting the fan for a while," said Scott. "And I've got a few footprints up my back. But we did it."

"It was a great partnership," said Rick. "Real team spirit."

I looked around at the rest of the zoo people. The administrators huddled together in one group, the curators and keepers in another. The administrators were excited, while the animal personnel were somber. The burden of the new exhibit had shifted, and now the curators and keepers had to carry the load. They had to help the animals get adjusted to the new environment. Roseann watched from the railing of one breezeway. Her mouth was pinched and her eyes were liquid, as though she were about to cry.

"Roseann's very emotionally involved right now," Bob Berghaier explained. "But she'll relax when the animals get used to the new place. That may take a couple of months, but they'll come around."

I asked him how he was doing. "Okay," he said. "I worry about John a little. He scared me when I first saw him down there. I'd never seen him without bars or glass. He stared at me really fierce. He was grinding his teeth and tightening his lips like he used to. He hadn't done that in a couple of months. And I can't condition him down here the way I used to because I can't get the reward to him as quickly. So we'll just have to see."

But John has problems of his own. "Samantha's starting to act independent," said Berghaier. "She's starting to stray— go off on her own to the far side of the exhibit. Independent females do that in the wild. They drift off and join other troops. Pretty soon, John's going to whack her to keep her in line."

Rick Biddle and Scott Schultz came over to look at the sculpture of Massa. Over the weekend, the zoo was planning to plant a time capsule under the statue. The capsule would contain a lock of Bong's hair, the several thousand "Good Deeds"—titles to square inches of the primate center—that Scott had marketed to zoo patrons, and some other relics. The bronze shone brightly on Massa's sagittal crest and heavy brow where so many zoo visitors had patted the sculpture. When Rick himself pressed a hand down on Massa's head, he noticed that the sculpture rocked on its foundation. "Hey," he said, "this thing is loose!" Scott looked at it. The two jostled it back and forth. They summoned a maintenance man, who used his pliers to tighten the bolts. Rick tested the sculpture once more, and this time Massa held firm. "There," he said.